Innovations in Software Engineering for Defense Systems

Oversight Committee for the Workshop on Statistical Methods
in Software Engineering for Defense Systems

Siddhartha R. Dalal, Jesse H. Poore, and
Michael L. Cohen, editors

Committee on National Statistics
Division of Behavioral and Social Sciences and Education

and

Committee on Applied and Theoretical Statistics
Division on Engineering and Physical Sciences

NATIONAL RESEARCH COUNCIL
OF THE NATIONAL ACADEMIES

THE NATIONAL ACADEMIES PRESS
Washington, D.C.
www.nap.edu

THE NATIONAL ACADEMIES PRESS • 500 Fifth Street, NW • Washington, DC 20001

NOTICE: The project that is the subject of this report was approved by the Governing Board of the National Research Council, whose members are drawn from the councils of the National Academy of Sciences, the National Academy of Engineering, and the Institute of Medicine. The members of the committee responsible for the report were chosen for their special competences and with regard for appropriate balance.

This material is based upon work supported by the National Science Foundation under Grant No. SBR-9709489. Any opinions, findings, conclusions, or recommendations expressed in this publication are those of the author(s) and do not necessarily reflect the views of the National Science Foundation.

International Standard Book Number 0-309-08983-2 (Book)
International Standard Book Number 0-309-52631-0 (PDF)
Library of Congress Control Number 2003110750

Additional copies of this report are available from National Academies Press, 500 Fifth Street, N.W., Lockbox 285, Washington, DC 20055; (800) 624-6242 or (202) 334-3313 (in the Washington metropolitan area); Internet, http://www.nap.edu.

Printed in the United States of America.

Copyright 2003 by the National Academy of Sciences. All rights reserved.

Suggested citation: National Research Council. (2003). *Innovations in Software Engineering for Defense Systems*. Oversight Committee for the Workshop on Statistical Methods in Software Engineering for Defense Systems. S.R. Dalal, J.H. Poore, and M.L. Cohen, eds. Committee on National Statistics, Division of Behavioral and Social Sciences and Education. Washington, DC: The National Academies Press.

THE NATIONAL ACADEMIES
Advisers to the Nation on Science, Engineering, and Medicine

The **National Academy of Sciences** is a private, nonprofit, self-perpetuating society of distinguished scholars engaged in scientific and engineering research, dedicated to the furtherance of science and technology and to their use for the general welfare. Upon the authority of the charter granted to it by the Congress in 1863, the Academy has a mandate that requires it to advise the federal government on scientific and technical matters. Dr. Bruce M. Alberts is president of the National Academy of Sciences.

The **National Academy of Engineering** was established in 1964, under the charter of the National Academy of Sciences, as a parallel organization of outstanding engineers. It is autonomous in its administration and in the selection of its members, sharing with the National Academy of Sciences the responsibility for advising the federal government. The National Academy of Engineering also sponsors engineering programs aimed at meeting national needs, encourages education and research, and recognizes the superior achievements of engineers. Dr. Wm. A. Wulf is president of the National Academy of Engineering.

The **Institute of Medicine** was established in 1970 by the National Academy of Sciences to secure the services of eminent members of appropriate professions in the examination of policy matters pertaining to the health of the public. The Institute acts under the responsibility given to the National Academy of Sciences by its congressional charter to be an adviser to the federal government and, upon its own initiative, to identify issues of medical care, research, and education. Dr. Harvey V. Fineberg is president of the Institute of Medicine.

The **National Research Council** was organized by the National Academy of Sciences in 1916 to associate the broad community of science and technology with the Academy's purposes of furthering knowledge and advising the federal government. Functioning in accordance with general policies determined by the Academy, the Council has become the principal operating agency of both the National Academy of Sciences and the National Academy of Engineering in providing services to the government, the public, and the scientific and engineering communities. The Council is administered jointly by both Academies and the Institute of Medicine. Dr. Bruce M. Alberts and Dr. Wm. A. Wulf are chair and vice chair, respectively, of the National Research Council.

www.national-academies.org

OVERSIGHT COMMITTEE FOR THE WORKSHOP ON STATISTICAL METHODS IN SOFTWARE ENGINEERING FOR DEFENSE SYSTEMS

DARYL PREGIBON *(Chair),* AT&T Laboratories, Florham Park, New Jersey
BARRY BOEHM, Computer Science Department, University of Southern California
SIDDHARTHA R. DALAL, Xerox Corporation, Webster, New York
WILLIAM F. EDDY, Department of Statistics, Carnegie Mellon University
JESSE H. POORE, Department of Computer Science, University of Tennessee, Knoxville
JOHN E. ROLPH, Marshall School of Business, University of Southern California
FRANCISCO J. SAMANIEGO, Division of Statistics, University of California, Davis
ELAINE WEYUKER, AT&T Laboratories, Florham Park, New Jersey

MICHAEL L. COHEN, *Study Director*
MICHAEL SIRI, *Project Assistant*

COMMITTEE ON NATIONAL STATISTICS
2002-2003

JOHN E. ROLPH *(Chair)*, Marshall School of Business, University of Southern California
JOSEPH G. ALTONJI, Department of Economics, Yale University
ROBERT M. BELL, AT&T Laboratories, Florham Park, New Jersey
LAWRENCE D. BROWN, Department of Statistics, University of Pennsylvania
ROBERT M. GROVES, Survey Research Center, University of Michigan
JOEL L. HOROWITZ, Department of Economics, Northwestern University
WILLIAM D. KALSBEEK, Department of Biostatistics, University of North Carolina
ARLEEN LEIBOWITZ, School of Public Policy Research, University of California, Los Angeles
THOMAS A. LOUIS, Department of Biostatistics, Johns Hopkins University
VIJAYAN NAIR, Department of Statistics, University of Michigan
DARYL PREGIBON, AT&T Laboratories, Florham Park, New Jersey
KENNETH PREWITT, School of International and Public Affairs, Columbia University
NORA CATE SCHAEFFER, Department of Sociology, University of Wisconsin, Madison
MATTHEW D. SHAPIRO, Department of Economics, University of Michigan

ANDREW A. WHITE, *Director*

COMMITTEE ON APPLIED AND THEORETICAL STATISTICS
2002-2003

SALLIE KELLER-McNULTY *(Chair)*, Los Alamos National Laboratory
SIDDHARTHA DALAL, Xerox Corporation, Webster, New York
CATHRYN DIPPO, Consultant, Washington, DC
THOMAS KEPLER, Duke University Medical Center, Durham, North Carolina
DANIEL L. McFADDEN, University of California, Berkeley
RICHARD OLSHEN, Stanford University
DAVID SCOTT, Rice University, Houston, Texas
DANIEL L. SOLOMON, North Carolina State University
EDWARD C. WAYMIRE, Oregon State University, Corvallis, Oregon
LELAND WILKINSON, SPSS, Incorporated, Chicago

SCOTT T. WEIDMAN, *Director*

Acknowledgments

The Workshop on Statistical Methods in Software Engineering for Defense Systems, jointly sponsored by the Committee on National Statistics and the Committee on Applied and Theoretical Statistics of the National Research Council (NRC), grew out of a need to more fully examine issues raised in the NRC reports *Statistics, Testing, and Defense Acquisition: New Approaches and Methodological Improvements* (NRC, 1998) and *Reliability Issues for Defense Systems: Report of a Workshop* (NRC, 2002). The former stressed, among other things, the importance of software requirements specification, and the latter demonstrated the benefits gained from model-based testing. Concurrent with these activities, there was an increasing recognition in the Office of the Secretary of Defense that software problems were a major source of defects, delays, and cost overruns in defense systems in development, and that approaches to addressing these problems were very welcome.

The Workshop on Statistical Methods in Software Engineering for Defense Systems, which took place July 19-20, 2001, was funded jointly by the offices of the Director of Operational Test and Evaluation (DOT&E) and the Undersecretary of Defense for Acquisition, Technology, and Logistics (OUSD(AT&L)). DOT&E, headed by Tom Christie, was ably represented at various planning meetings by Ernest Seglie, the science advisor for DOT&E, and OUSD (AT&L) was represented by Nancy Spruill. Both Ernest Seglie and Nancy Spruill were enormously helpful in suggesting topics for study, individuals to serve as expert defense discussants, and invited guests to supplement floor discussion. Their contribution was essential to the success of this workshop.

We would also like to thank the following Office of the Secretary of Defense employees for their assistance: Walter Benesch, Michael Falat, Jack Ferguson, Ken Hong Fong, Liz Rodriguez Johnson, Steven King, Bob Nemetz, Rich Turner, and Bob Williams. We are very grateful to Norton Compton for his help on a number of administrative issues and his can-do spirit. Scott Weidman, staff director of the NRC's Committee on Applied and Theoretical Statistics, provided very helpful advice in structuring the workshop. We would also like to thank Art Fries of the Institute for Defense Analysis, Dave Zubrow of the Software Engineering Institute, and David Nicholls of IIT Research Institute for their assistance.

The Oversight Committee for the Workshop on Statistical Methods in Software Engineering for Defense Systems met and interacted only informally, often via telephone and e-mail, and in person in small groups of two and three. It relied to a great extent on the content of the presentations given at the workshop. Therefore, the Committee would like to express its gratitude to the speakers and the discussants: Frank Apicella, Army Evaluation Center, Vic Basili, University of Maryland, Patrick Carnes, Air Force Operational Test and Evaluation Center, Ram Chillarege, consultant, Tom Christian, WR-ALC, Delores Etter, Office of the Secretary of Defense, William Farr, Naval Surface Warfare Center, Jack Ferguson, Office of the Undersecretary of Defense for Science and Technology, Janet Gill, PAX River Naval Air Systems, Software Safety, Constance Heitmeyer, Naval Research Lab, Brian Hicks, Joint Advanced Strike Technology, Mike Houghtaling, IBM, Jerry Huller, Raytheon, Ashish Jain, Telcordia Technologies, Stuart Knoll, Joint Staff, J-8, Scott Lucero, Army SW Metrics, Army Evaluation Center, Luqi, Naval Postgraduate School, Ron Manning, Raytheon, David McClung, University of Texas, Margaret Myers, Office of the Assistant Secretary of Defense for Command, Control, Communications and Intelligence, Jim O'Bryon, Office of the Director of Operational Test and Evaluation, Lloyd Pickering, Army Evaluation Center, Stacy J. Prowell, University of Tennessee, Manish Rathi, Telcordia Technologies, Harry Robinson, Microsoft, Mike Saboe, Tank Automotive and Armaments Command, Ernest Seglie, Office of the Director of Operational Test and Evaluation, Linda Sheckler, Penn State University, Applied Research Lab, Kishor Trivedi, Duke University, Amjad Umar, Telcordia Technologies, and Steven Whitehead, Operational Test and Evaluation Force Technical Advisor. Further, we would like to give special thanks to Constance Heitmeyer and Amjad Umar for their help in writing sections of the report.

We would also like to thank Michael Cohen for help in organizing the workshop and in drafting sections of the report, and Michael Siri for his excellent attention to all administrative details. We would also like to thank Cameron Fletcher for meticulous technical editing of the draft report, catching numerous errors.

This report has been reviewed in draft form by individuals chosen for their diverse perspectives and technical expertise, in accordance with procedures approved by the Report Review Committee of the NRC. The purpose of this independent review is to provide candid and critical comments that will assist the institution in making the published report as sound as possible and to ensure that the report meets institutional standards for objectivity, evidence, and responsiveness to the study charge. The review comments and draft manuscript remain confidential to protect the integrity of the deliberative process.

We thank the following individuals for their participation in the review of this report: Robert M. Bell, AT&T Research Laboratories, Florham Park, NJ, Philip E. Coyle, III, Consultant, Los Angeles, CA, Michael A. Houghtaling, IBM/ Systems Group/ Storage Systems Division, Tucson, AZ, and Lawrence Markosian, QSS Group, Inc., NASA Ames Research Center, Moffett Field, CA. Although the reviewers listed above have provided many constructive comments and suggestions, they were not asked to endorse the conclusions or recommendations nor did they see the final draft of the report before its release. The review of this report was overseen by William G. Howard, Jr., Independent Consultant, Scottsdale, AZ. Appointed by the NRC, he was responsible for making certain that an independent examination of this report was carried out in accordance with institutional procedures and that all review comments were carefully considered. Responsibility for the final content of this report rests entirely with the authoring committee and the institution.

        Daryl Pregibon, *Chair*
        Workshop on Statistical Methods in Software
        Engineering for Defense Systems

Contents

EXECUTIVE SUMMARY 1

1 MOTIVATION FOR AND STRUCTURE OF THE
  WORKSHOP 5
  Structure of the Workshop, 6
  Workshop Limitations, 10

2 REQUIREMENTS AND SOFTWARE ARCHITECTURAL
  ANALYSIS 12
  Software Cost Reduction, 13
  Sequence-Based Software Specification, 16

3 TESTING METHODS AND RELATED ISSUES 20
  Introduction to Model-Based Testing, 20
  Markov Chain Usage Models, 24
  AETG Testing, 29
  Integration of AETG and Markov Chain Usage Models, 32
  Test Automation, 33
  Methods for Testing Interoperability, 34

| 4 | DATA ANALYSIS TO ASSESS PERFORMANCE AND TO SUPPORT SOFTWARE IMPROVEMENT | 40 |

Measuring Software Risk, 40
Fault-Tolerant Software: Measuring Software Aging and
   Rejuvenation, 44
Defect Classification and Analysis, 45
Bayesian Integration of Project Data and Expert Judgment in
   Parametric Software Cost Estimation Models, 48

| 5 | NEXT STEPS | 51 |

| REFERENCES | 56 |

APPENDICES

A	WORKSHOP AGENDA AND SPEAKERS	61
B	GLOSSARY AND ACRONYM LIST	66
C	BIOGRAPHICAL SKETCHES	70

Executive Summary

Recent rough estimates are that the U.S. Department of Defense (DoD) spends at least $38 billion a year on the research, development, testing, and evaluation of new defense systems; approximately 40 percent of that cost—at least $16 billion—is spent on software development and testing. There is widespread understanding within DoD that the effectiveness of software-intensive defense systems is often hampered by low-quality software as well as increased costs and late delivery of software components. Given the costs involved, even relatively incremental improvements to the software development process for defense systems could represent a large savings in funds. And given the importance of producing defense software that will carry out its intended function, relatively small improvements to the quality of defense software systems would be extremely important to identify.

DoD software engineers and test and evaluation officials may not be fully aware of a range of available techniques, because of both the recent development of these techniques and their origination from an orientation somewhat removed from software engineering, i.e., from a statistical perspective.

The panel's charge therefore was to convene a workshop to identify statistical software engineering techniques that could have applicability to DoD systems in development. It was in response both to previous reports that identified a need for more attention to software and to the emergence of techniques that have been widely applied in industry and have demonstrated impressive benefits in the reduction of errors as well as testing costs and time.

The panel structured its task consistent with a statistical approach to collecting information on system development. This approach has three stages: (1) determination of complete, consistent, and correct requirements; (2) selection of an effective experimental design to guide data collection on system performance; and (3) evaluation of system performance, consistent with requirements, using the data collected. The intent was to focus on methods that had been found useful in industrial applications. Some of the issues selected for study were suggested by DoD software experts for specific areas to address that would make the workshop relevant to their concerns and needs.

The panel's work was not intended to be comprehensive in identifying techniques that could be of use to DoD in software system development and for software engineering, or even for statistically oriented software engineering. Without question, a number of very effective techniques in software engineering were not represented at the workshop nor considered by the panel.

The workshop sessions on specification of requirements focused on software cost reduction and sequence-based software specification. The sessions on test design addressed Markov chain usage testing and combinatorial design testing. Presentations on software evaluation discussed risk assessment, cost modeling, models of software process improvement, and fault-tolerant software. The presentations on software evaluation suggested the great variety of issues that can be addressed through data analysis and, therefore, the value of collecting and analyzing data on system performance, both during system development and after fielding a system.

After the workshop, the oversight committee interacted mainly by e-mail and, to a lesser extent, by telephone and in person in small groups of two or three. Informed and primarily motivated by the workshop's presentations, but augmented by these subsequent interactions, the oversight committee decided to issue recommendations for further efforts in this area.

> **Recommendation 1: Given the current lack of implementation of state-of-the-art methods in software engineering in the service test agencies, initial steps should be taken to develop access to—either in-house or in a closely affiliated relationship—state-of-the-art software engineering expertise in the operational or developmental service test agencies.**

Such expertise could be acquired in part in several ways, especially including internships for doctoral students and postdoctorates at the test and evaluation agencies, and with sabbaticals for test and evaluation agency staff at industry sites at which state-of-the-art techniques are developed and used.

> **Recommendation 2: Each service's operational or developmental test agency should routinely collect and archive data on software performance, including test performance data and data on field performance. The data should include fault types, fault times and frequencies, turnaround rate, use scenarios, and root cause analysis. Also, software acquisition contracts should include requirements to collect such data.**

> **Recommendation 3: Each service's operational or developmental test agency should evaluate the advantages of the use of state-of-the-art procedures to check the specification of requirements for a relatively complex defense software-intensive system.**

One effective way of carrying out this evaluation would be to develop specifications in parallel, using the service's current procedures, so that quality metric comparisons can be made. Each service would select a software system that: (1) requires field configuration, (2) has the capabilities to adapt and evolve to serve future needs, and (3) has both hardware and software that come from multiple sources.

> **Recommendation 4: Each service's operational or developmental test agency should undertake a pilot study in which two or more testing methods, including one model-based technique, are used in parallel for several software-intensive systems throughout development to determine the amount of training required, the time needed for testing, and the method's effectiveness in identifying software defects. This study should be initiated early in system development.**

Recommendations 3 and 4 propose that each of the services select software-intensive systems as case studies to evaluate new techniques. By doing so, test and evaluation officials can judge both the costs of training engineers to use the new techniques—the "fixed costs" of widespread implementation—and the associated benefits from their use in the DoD acquisition environment. Knowledge of these costs and benefits will aid DoD in deciding whether to implement these techniques more broadly.

There are insufficient funds in individual programs or systems under development to support the actions we recommend. Thus, support will be needed from the services themselves or from the Department of Defense.

Recommendation 5: DoD should allocate general research and development funds to support pilot and demonstration projects of the sort recommended in this report in order to identify methods in software engineering that are effective for defense software systems in development.

The panel notes that constraints hinder DoD from imposing on its contractors specific state-of-the-art techniques in software engineering and development that are external to the technical considerations of the costs and benefits of the implementation of the techniques themselves. However, the restrictions do not preclude the imposition of a framework, such as the capability maturity model.

Recommendation 6: DoD needs to examine the advantages and disadvantages of the use of methods for obligating software developers under contract to DoD to use state-of-the-art methods for requirements analysis and software testing, in particular, and software engineering and development more generally.

The techniques discussed in this report are consistent with these conditions and constraints. We also note that many of the techniques described in this report are both system oriented and based on behavior and are therefore applicable to both hardware and software components, which is an important advantage for DoD systems.

1

Motivation for and Structure of the Workshop

Recent rough estimates are that the U.S. Department of Defense (DoD) spends at least $38 billion a year on the research, development, testing, and evaluation of new defense systems, and that approximately 40 percent of that cost—at least $16 billion—is spent on software development and testing (Ferguson, 2001; *Aerospace Daily*, 2003). Given the costs involved, even relatively incremental improvements to the software development process for defense systems could represent a large savings in defense funds, in addition to producing higher-quality defense systems. Therefore, a high priority needs to be accorded to the identification of software engineering methods that can be used to provide higher-quality software at reduced costs.

In addition to impacts on quality improvement and cost savings, software problems are known to cause delays in the delivery of new defense systems and to result in reduced functionality in comparison to initial specifications. More importantly, field failures can lead to mission failure and even loss of life. These are all important reasons to support broad-based investigations into various approaches to improve the process of engineering software for defense systems.

In opening remarks at the workshop, Delores Etter, Deputy Under Secretary for Defense (Science and Technology), described the complicated software systems embedded in the Comanche RAH-66 Helicopter, in the NAVSTAR global positioning system, in the AEGIS weapon system, and in the Predator Unmanned Aerial Vehicle. For example, for AEGIS, there are 1,200K lines of code for display, 385K lines of code for the test, 266K lines of code for the weapon, 110K lines of code for training, and 337K lines of code for the command and decision making. Complicated software systems are ubiquitous in defense systems today.

The Workshop on Statistical Methods in Software Engineering for Defense Systems grew out of the work of the Panel on Statistical Methods for Testing and Evaluating Defense Systems. This Committee on National Statistics panel, funded by the DoD Director of Operational Test and Evaluation (DOT&E), examined the use of statistical methods in several separate arenas related to the design and evaluation of operational tests, and, more broadly, the process used for the development of defense systems (NRC, 1998). The panel identified a number of areas of application in which the problem-solving approach of statistical science (in contrast to simple, omnibus techniques) could be used to help provide extremely useful information to support decisions on defense system development. Two of the arenas examined by this panel were testing software-intensive systems and testing software architecture (see NRC, 1998, Chapter 8).

Following this panel's recommendation to continue the examination of the applicability of statistical techniques to defense system development at a more detailed level, DOT&E, along with the Office of Acquisition, Technology, and Logistics of the Office of the Secretary of Defense, initiated a series of workshops to explore in greater depth issues raised in the different areas of focus of the National Research Council (NRC, 1998). The first workshop, held June 9-10, 2000, addressed the issue of reliability assessment (see NRC, 2002, for details). The second workshop, held July 19-20, 2001, and jointly organized with the NRC's Committee on Applied and Theoretical Statistics, dealt with the use of statistical methods for testing and evaluation of software-intensive systems and is the chief basis for this report.[1]

STRUCTURE OF THE WORKSHOP

The Workshop on Statistical Methods in Software Engineering for Defense Systems was structured to correspond roughly with the steps to

---

[1] For related work, see NRC (1996).

carry out a statistical assessment of the functioning of any industrial system, whether hardware or software. The steps of a statistical analysis are: (1) specification of requirements, (2) selection of an experimental design to efficiently select test cases to collect information on system performance in satisfying requirements, and (3) analysis of the resulting experimental data (to estimate performance, check compliance with requirements, etc.). This structure was used to organize the workshop presentations and we use the same structure to organize this workshop report.

The linear structure of this report is unfortunate in that it does not communicate well the overlapping aspects of many of the methods described, given that research in one area is often relevant to others. This is one of the justifications for optimism that tools will be developed in the very near future that would further combine, say, requirements specifications, testing, cost estimation, and risk assessment as methods emanating from a unified framework. For example, overlap can already be seen in the development of software tools that test software functionality; known as test oracles, these tools are needed for use with model-based testing and could be assisted by tools developed for requirements specification. Conversely, the graph on which model-based testing relies can be used to support requirements specification.

We now provide some detail on the topics examined at the workshop in these three areas of software engineering. (See also the workshop agenda in Appendix A.)

Specification of requirements and software testability: A requisite for data collection on system performance is the specification of requirements, which defines the successful functioning of a system. Requirements need to be examined with respect to: (1) the possibility of verification, (2) completeness, (3) consistency, (4) correctness, and (5) complexity. It is useful to judge how complicated a system is in order to estimate how many replications might be needed to adequately test a software system (given an agreed-upon definition of adequate). The complexity of software is dependent on its architecture, which is related to the specification of requirements. To address both whether requirements are well specified and whether the software is structured in a way to facilitate testing, the workshop included a session concerning specification of requirements for software systems.

Selection of test cases—experimental design: The better one understands software performance, the better one is able to predict the expected behavior of systems. The better one understands the software development process, the better one can intelligently allocate scarce resources (including personnel, tools, hardware, and machine cycles) to address any deficiencies. By collecting data and analyzing them, one can develop a better understanding of both software performance and potential deficiencies in the software development process. To collect data on how a software system operates, the system must be exercised on selected test scenarios, i.e., a suite of test inputs. It is sometimes believed, particularly with automated testing, that since many software systems execute in small fractions of a second, a software system could be executed a very large number of times without needing to consider any notions of optimal selection of test scenarios. However, inherent system complexity leads to an astronomical number of test inputs for virtually all defense software systems, hence the need to carefully select test inputs. Furthermore, many systems either do not execute quickly or else have long set-up times for test cases. For example, testing the interoperability of a system that is composed of various subsystems which can be represented by different versions or releases is typically not an automated activity and therefore necessitates the careful selection of test inputs. Thus concepts and tools from the statistical field of experimental design are relevant to consider.

Two statistically oriented approaches for the selection of inputs for software testing have been successfully applied in a variety of industrial settings. They are both examples of "model-based" testing methods, which rely (possibly implicitly) on a graphical representation of the software in action. The nodes of the graph represent user-relevant states of the software, and various user actions are represented as transitions from one node to another. One of the two approaches to input selection presented at the workshop relies on a Markov chain model of software use in transitioning through the graphical representation, described below. The second approach, referred to as combinatorial design, identifies a surprisingly small collection of software test inputs that includes all k-wise combinations of input fields (typically at a given node of the graphical representation), where k is typically chosen to be small, often 2 or 3. Both approaches have been used in industrial applications to help provide high-quality software with substantial cost savings and schedule reductions.

Given the different advantages of these two methods, there is interest in developing a hybrid method that combines and retains the advantages of both, and so some possibilities for achieving this are discussed. In addition, since Markov chain usage testing, while quite efficient, often requires a substantial test sample size to provide reliable statistics on software performance, test automation remains crucial, and so such methods were also briefly examined at the workshop.

A common problem that is a high priority for the defense acquisition community is that of software systems that are composed of subsystems that have versions that are collectively incompatible. (This is referred to as the interoperability problem.) Many software-intensive defense systems utilize commercial-off-the-shelf (COTS) systems as subroutines or as lower-level component systems. Different COTS systems have separate release schedules and, combined with the fact that any custom software systems included as components will also be subject to modifications, there is the possibility that some of these configurations and modifications may be incompatible. A session at the workshop provided a description of many of the tools used in industry to enhance interoperability, and showed that combinatorial design methods can be used to measure the extent of interoperability in a system of systems.

Analysis of test data: Data that are collected (possibly using the above experimental design procedures) consist of test inputs, which can be considered from a statistical perspective as "independent variables," and the associated results of the software system exercised on those inputs, which can be thought of as "dependent variables." These input-output pairs can be used both to help determine whether the software functioned successfully for specifically chosen inputs and to measure the extent of, and identify the sources of, defects in a software system. Furthermore, based on these data and the assessment of whether the software functioned properly, a variety of additional analyses can be carried out concerning the performance of the system and the software development process. Accordingly, a session at the workshop was organized to consider various uses of data collected on software performance and development. This session included presentations on measurement of software risk, software aging, defect analysis and classification, and estimation of the parameters of models predicting the costs of software development.

Two examples of performance measurement not covered at the workshop are reliability growth modeling and decision making on when to release a product. Dozens of models on reliability growth have been suggested in the literature with varying interpretations of the impact on system reliability associated with code changes. Some assumptions imply better performance after changes, while others imply worse performance. Some assumptions imply linear growth, while others imply exponential growth. The test data are used to fit the parameters of these models and then projections are made. This literature includes contributions by Musa, Goel, Littlewood, and Veral, among others. (For specific references to their work, see Lyu, 1996.) With respect to decisions on release dates for new software products, IBM uses test data not only to decide on the optimal release date but also to estimate field support budgets for the life of the product or until the next version is released (for details, see Whittaker and Agrawal, 1994). The literature on when to release a product is considerable; two recent contributions are Dalal and Mallows (1988) and Dalal and McIntosh (1994).

The workshop focused on methods that were already in use, for which mature tools existed, and that were generally believed to be readily adaptable to defense systems. Many of the techniques described in this report have already enjoyed successful application in DoD or DoD-like applications. For example, a Raytheon application of testing based on Markov chain usage models resulted in an 80 percent reduction in the cost of automated testing along with a substantial reduction in the percentage of project resources required for testing (17-30 percent versus 32-47 percent). Another Raytheon application of combinatorial design for a satellite control center experienced a 68 percent reduction in test duration and a 67 percent savings in test labor costs. And IBM Storage Systems Division reported on the first industrial use of model-based statistical testing for mass storage controllers. To help show the maturity of the methods, the workshop concluded with demonstrations of many of the software tools that had been described in the presentations.

WORKSHOP LIMITATIONS

This workshop was not designed to, and could not, address all of the recent innovations that are already implemented or under development as part of software engineering practice in industry. In no respect should the workshop or this report be considered a comprehensive examination of tools for software engineering. The purpose was to examine some promising statistically oriented methods in software engineering, necessarily omitting many important techniques. The workshop presentations can therefore be used to support only relatively broad recommendations concerning next steps for DoD in its adoption of software engineering methods.

Some topics that might be covered in a follow-up workshop of this type are design for testability, linking specifications requirements to testing models, development of certification testing protocols, and the behavior of systems of systems. The latter topic would entail a much broader look into issues raised by the development of systems of systems, rather than the narrow problem of testing for interoperability, which is mainly to identify faults. Another topic that could be covered is the selection and application of appropriate testing tools and strategies to achieve specific quality assurance requirements (as opposed to functional specification requirements for the software system). This topic could be examined more generally in terms of software process simulation models, which have obtained some use in industry and can be applied to optimize testing strategies as well as other aspects of the software development life cycle (for example, see Raffo and Kellner, 1999).

In the remainder of this report, the methods presented at the workshop for possible use by DoD are described. In addition, some recommendations are included relative to the implementation of these and related methods of software engineering practice for use with software-intensive defense systems. The report is organized as follows. This introduction is followed by chapters on requirements analysis, model-based testing methods and methods to address interoperability, and performance and data analysis to support software evaluation and improvement of software development practice. The report concludes with a short summary chapter containing recommendations for next steps.

2

Requirements and Software Architectural Analysis

Because specifications of software requirements are typically expressed in "natural" rather than technical language, they suffer from ambiguity and incompleteness. Poorly specified requirements are often the cause of software defects. It is believed that roughly half of software defects are introduced during the requirements and functional analysis stage,[1] and some well-known software development projects have had a large percentage of serious software defects introduced during the specification of requirements. Therefore, utilization of techniques that improve the software requirements can save substantial time and money in software development and will often result in higher-quality software. The ability to develop high-quality requirements specifications is of crucial importance to the Department of Defense (DoD) in particular, since catastrophic failure (including loss of life) can result from defects in some DoD software-intensive systems. Clearly, any important advances in methods for requirements specification and checking need to be considered for use with DoD systems.

In addition to the more timely identification of software defects and design problems, the development of better requirements specifications can

---

[1] Results of a 1992 study (Lutz, 1993) of defects detected during the integration and testing of the Voyager and Galileo spacecraft showed that 194 of 197 significant software defects could be traced to problems in the specification of function and interface requirements.

also facilitate software testing by establishing a solid foundation for model-based testing and for estimating the testability of a software system.

In this section we discuss in detail two methodologies useful for requirements specification: software cost reduction and sequence-based specification.

SOFTWARE COST REDUCTION

Constance Heitmeyer of the Naval Research Laboratory described software cost reduction (SCR), a set of techniques for developing software, pioneered by David Parnas and researchers from the Naval Research Laboratory (NRL) beginning in 1978. A major goal of the early research was to evaluate the utility and scalability of software engineering principles by applying them to the reconstruction of software for a practical system. The system selected was the Operational Flight Program (OFP) for the U.S. Navy's A-7 aircraft. The process of applying the principles to the A-7 requirements produced new techniques for devising precise, unambiguous requirements specifications. These techniques were demonstrated in a requirements document for the A-7 OFP (Heninger, 1980). Further research by Heitmeyer and her group during the 1990s produced a formal model to define the special SCR notation (Heitmeyer, Jeffords, and Labaw, 1996) and a suite of software tools for verifying and validating the correctness of SCR-style requirements specifications (Heitmeyer, Kirby, Labaw, and Bharadwaj, 1998).

A-7 and Later Developments

The A-7 requirements document, a complete specification of the required behavior of the A-7 OFP, demonstrated the SCR techniques for specifying software requirements (Heninger, 1980). It introduced three major aspects of SCR: the focus on system outputs, a special tabular notation for specifying each output, and a set of criteria for evaluating a requirements document. A critical step in constructing an SCR software requirements document is to identify all outputs that the software must produce and to express the value of each output as a mathematical function of the state and history of the environment. To represent these functions accurately, unambiguously, and concisely, the A-7 document introduced a special tabular notation that facilitates writing and understanding the functions and also aids in detecting specification errors, such as missing cases and ambiguity. A requirements specification must satisfy three important criteria in order to be acceptable: (1) completeness (i.e., any implementation satisfying every statement in the requirements document should be acceptable), (2) freedom from implementation bias (i.e., requirements should not address how the requirements are addressed), and (3) organization as a reference document (i.e., information in the document should be easy to find).

During the 1980s and 1990s, a number of organizations in both industry and government (such as Grumman, Bell Laboratories, NRL, and Ontario Hydro) used SCR to document the requirements of a wide range of practical systems. These systems include the OFP for the A-6 aircraft (Meyers and White, 1983), the Bell telephone network (Hester, Parnas, and Utter, 1981), and safety-critical components of the Darlington nuclear power plant (Parnas, Asmis, and Madey, 1991). In 1994, SCR was used to document the requirements of the operational flight program of Lockheed's C-130J aircraft (Faulk, 1995). The Lockheed requirements document contains over 1,000 tables and the operational flight program over 250K lines of Ada source code, thus demonstrating that SCR scales.

Until the mid-1990s, SCR requirements specifications were analyzed for defects using manual inspection. While such inspection can expose many defects in specifications, it has two serious shortcomings. First, it can be very expensive. In the certification of the Darlington system, for example, the manual inspection of SCR tabular specifications cost millions of dollars. Second, human inspections often overlook defects. For example, in a 1996 study by NRL, mechanized analysis of tables in the A-7 requirements document exposed 17 missing cases and 57 instances of ambiguity (Heitmeyer, Jeffords, and Labaw, 1996). These flaws were detected after the document had been inspected by two independent review teams. Thus, while human effort is critical to creating specifications and manual inspections can detect many specification errors, developing high-quality requirements specifications in industrial settings requires automated tool support. Not only can such tools find specification errors that manual inspections miss, they can do so more cheaply.

To establish a formal foundation for tools supporting the development of an SCR requirements specification, Heitmeyer and her research group formulated a formal model to rigorously define the implicit state machine model that underlies an SCR requirements specification. In the model, the system, represented as a state machine, responds to changes in the value of environmental quantities that it monitors (represented as monitored variables) by changing state and possibly by changing the values of one or more environmental quantities that it controls (represented as controlled variables). For example, an avionics system might be required to respond to a severe change in air pressure by sounding an alarm; the change in pressure corresponds to a change in a monitored quantity, while the sounding of the alarm represents a change in a controlled quantity.

NRL has designed a suite of software tools for constructing and analyzing SCR requirements specifications. The SCR tools have been designed to support a four-step process for developing requirements specifications. First, the user constructs a requirements specification using the SCR tabular notation. Second, the user invokes a consistency checker (Heitmeyer, Jeffords, and Labaw, 1996) to automatically detect syntax and type defects, missing cases, ambiguity, and circular definitions in the specification. When a defect is detected, the consistency checker provides detailed feedback to aid in defect correction. Third, to validate the specification, the user may run *scenarios* (sequences of changes in the monitored quantities) through the SCR simulator (Heitmeyer, Kirby, Labaw, and Bharadwaj, 1998) and analyze the results to ensure that the specification captures the intended behavior. To facilitate validation of the specification by application experts, the simulator supports the rapid construction of graphical user interfaces, customized for particular application domains. Finally, the user may analyze the requirements specification for critical application properties, such as security and safety properties, using static analysis tools. For example, the user may run a model checker, such as SPIN (Holtzman, 1997), to detect property violations, or a verification tool, such as the Timed Automata Modeling Environment (TAME) (Archer, 1999), to verify properties. In using TAME, a specialized interface to the general-purpose theorem-prover prototype verification system, completing a proof may require proving auxiliary properties. To construct these auxiliary properties, the user may invoke the SCR invariant generator (Jeffords and Heitmeyer, 1998), a tool that automatically constructs *state invariants* (properties true of every reachable state) from an SCR requirements specification.

Industrial Examples

Three specific case studies were presented at the workshop. In the first, engineers at Rockwell Aviation applied the SCR tools to a specification of their Flight Guidance System. Despite extensive prior manual reviews of the specification, the tools discovered 28 defects, many of them serious. The second case study involved a specification developed by Electric Boat of the software requirements of a Torpedo Tube Control Panel (TTCP). The software in this case was approximately 15,000 lines of relatively complex code. In under five person-weeks, the contractor specification was translated into SCR, and the SCR tools were used to analyze the specification for errors and to develop a realistic simulator for demonstrating and validating the required behavior of the TTCP. Analysis with the SCR tools and the SPIN model checker identified a serious safety violation in the SCR specification. In the third case study, an SCR requirements specification was devised for a software-based cryptographic device under development at NRL. TAME was used to analyze the specification for eight security properties (e.g., if a cryptographic device is tampered with, the keys are "zero-ized"). The analysis showed that the SCR specification of the required behavior of the device satisfied seven of the properties and violated the eighth, and therefore the requirements were inconsistent.

SEQUENCE-BASED SOFTWARE SPECIFICATION

Another approach to the development and verification of complete, consistent, and correct specification of requirements is sequence-based specification, which is described in Prowell et al. (1999). The methodology was effectively illustrated with the example of a security alarm system. The method begins with stage I, the "black box" definition. The first step of this stage is to "tag" the requirements, i.e., give each of the initially specified requirements a numerical tag (from 1 to n). In the example of the security alarm, the fourth requirement is that if a trip signal occurs while the security alarm is set, a highly pitched tone (alarm) is emitted. Through the remainder of the sequence-based specification process, it is likely that additional requirements will be identified, and they are referred to as derived requirements and are typically given a numerical tag preceded by a "D." The tagging of requirements is followed by a listing of the input stimuli that the system is expected to respond to and a listing of the responses that the system is expected to provide, with both linked back to the requirements. For example, stimuli include setting the alarm, tripping the alarm, entering a bad digit in sequence for turning off the alarm, and entering a good digit in sequence for turning off the alarm. An example of a response is "alarm is turned on when it is tripped," traced back to the fourth requirement.

All possible sequences of input stimuli are then listed in strict order, starting with stimuli of length zero, then one, and so on. For example, if "S" represents setting the alarm, "T" tripping the alarm, "B" entering a bad digit, and "G" entering a good digit, then "SGGT" is a stimuli sequence of length four. Enumeration of input sequences ends when all sequences of a given length are either "illegal" or equivalent to a previous sequence, where equivalent means that the two sequences stimulate identical behavior of the system. The correct system response linked to each stimuli sequence is determined by tracing back to the associated requirement tag or tags. When no requirements exist to determine the response, derived requirements are defined and tagged. This process ensures that system behavior is defined for all possible (finite-length) input sequences (an infinite number), and the enumeration process guarantees that this will be done while considering the smallest possible number of sequences.

When the enumeration is complete, each stimulus sequence has been mapped to a response and one or more requirements. If each requirement has been cited at least once the specification is both complete and—since each stimuli sequence is mapped to a single response—consistent. Finally, engineers (or domain experts) can determine whether each derived (and original) requirement is correct.

The complete enumeration provides the starting point for stage II, the "state box" definition. State variables are used to distinguish between different states of the system. Each nonequivalent stimuli sequence is examined to identify the unique conditions in the sequence, and state variables are invented to represent these conditions. In the example of the security alarm, the states are whether the device is on or off, whether the alarm is on or off, and whether the digits entered are consistent with the password. Whereas the black box representation of the system was expressed in terms of stimuli sequences and responses, the state box representation is expressed in terms of the current stimulus, the current state of the system, and the system response and state update. The correctness of the specifications can now be examined sequentially as stimuli are processed by the system. (This state box is an excellent starting point for the model underlying Markov chain usage testing, discussed in the following section.)

Sequence-based specification provides a constructive process for defining the state machine in a manner that ensures complete, consistent, and correct requirements, whereas SCR, in attaining the same goals, relies on intuition assisted by graphical tools to develop the state machine. A schema called Z (developed by Woodcock and others; see, e.g., Woodcock and

Davies, 1996) is a type of "set theory" for the formal prescription of state machines. There are tools available for Z and some impressive applications have been made by those mentioned above in collaboration with IBM's Cambridge Lab. A related method developed at IBM's Vienna Lab and known as VDM (for Vienna Development Method) is less formal than Z. Other approaches include that of Luckham and others (see, e.g., Luckham et al., 1995) who have developed a system that is fully formal, and SRI, whose formal system was developed by Rushby and others. Both of these approaches have tools that assist in their application. In the telecommunications industry, there are two state-based systems, Statemate[2] and State Description Language, neither of which is constructive but both of which are based on a well-understood application and which use idioms of the application that are accepted as correct based on extensive engineering experience and field success. David Parnas (see Bartoussek and Parnas, 1978, and Parnas and Wang, 1989) has continued SCR in a direction using relations rather than functions and developed table templates for various standard expressions of common needs in software. Prototype tools have also been developed to convert tabular representations of relations to code (see e.g., Leonard and Heitmeyer, 2003).

In conclusion, the specification of requirements, if poorly implemented, can be an important source of software defects for defense systems. A number of relatively new approaches to checking the specification of requirements have now been developed that can be used to create a set of consistent, complete, and correct requirements. Two leading approaches were described at the workshop, SCR and sequence-based software specification. SCR has been successfully applied to a number of defense and defense-related software systems. Sequence-based specification has some theoretical advantages given that it is more algorithmic in its application. Both methods are still relatively new and are being further developed, and software tools are increasingly available to facilitate their application. In

---

[2]The semantics of Statemate is being applied to the definition of formal behavioral semantics for the graphical state diagram notation employed by the Unified Modeling Language standard. This modeling and domain analysis notation for system specifications is supported by many vendor tools. See, e.g., Harel and Kupferman (2002) and Harel (2001).

addition, other related approaches have similar advantages. The workshop participants generally agreed on the potential benefit of applying these methods widely to defense software systems in development, and therefore the value of continued testing and analysis of the applicability of such methods to defense systems.

3

Testing Methods and Related Issues

Testing of software in development has two primary purposes. First, testing methods are used to assess the quality of a software system. Second, while practitioners stress that it is not a good idea to try to "test quality into" a deficient software system since some defects are likely unavoidable, testing is very important for discovering those that have occurred during software development. So for both verifying the quality of software and identifying defects in code, testing is a vital component of software engineering. But software testing is expensive, typically requiring from one-fifth to one-third of the total development budget (see Humphrey, 1989), and can be time-consuming. As a result, methods that would improve the quality and productivity of testing are very important to identify and implement. Because testing is an area in which statistically oriented methods are having an important impact on industrial software engineering, the workshop included a session on those methods.

INTRODUCTION TO MODEL-BASED TESTING

Two of the leading statistically oriented methods for the selection of inputs for software testing are forms of model-based testing, and therefore it is useful to provide a quick overview of this general approach.

Prior to the development of model-based testing, the general approaches used for software testing included manual testing, scripted automation, use of random inputs, finite state models, and production grammar models. Each of these approaches has serious deficiencies. For example, manual testing and scripted automation are extremely time-consuming since they require frequent updating (if they are not updated, they are likely to catch only defects that have already been identified). Random input models are very wasteful in that most of the inputs fail to "make sense" to the software and are immediately rejected. Finite state models, which operate at the same level of detail as the code, need to be extremely detailed in order to support effective testing.

The idea behind model-based testing is to represent the functioning of a software system using a model composed of user scenarios. Models can be implemented by a variety of structured descriptions, for example, as various tours through a graph. From an analytical and visual perspective it is often beneficial to represent the model as a structured tree or a graph. The graph created for this purpose uses nodes to represent observable, user-relevant states (e.g., the performance of a computation or the opening of a file). The arcs between a set of nodes represent the result of user-supplied actions or inputs (e.g., user-supplied answers to yes/no questions or other key strokes, or mouse clicks in various regions of the desktop) that correspond to the functioning of the software in proceeding from one state of use to another. (The graph produced for this purpose is at a much higher level and not code based, as opposed to that for a finite state model.)[1]

It should be pointed out before proceeding that even if every path were tested and no defects found, this would not guarantee that the software system was defect-free since there is a many-to-one relationship of inputs to paths through the graph. Furthermore, even if there are no logical errors in the software, a software system can fail due to a number of environmental factors such as problems with an operating system and erroneous inputs. Thus, to be comprehensive, testing needs to incorporate scenarios that involve these environmental factors.

To identify defects, one could consider testing every path through the graph. For all but the simplest graphical models, however, a test of every path would be prohibitively time-consuming. One feasible alternative is to efficiently choose test inputs with associated graphical paths that collectively contain every arc between nodes in the graphical model. There are algorithms that carry out this test plan; random walks through the graph

---

[1] One side benefit of developing a graphical model of the functioning of a software system is that it helps to identify ambiguities in the specifications.

are used, and more complicated alternatives are also available (see Gross and Yellen, 1998, for details). It is important to point out that the graph produced for this purpose is assumed to be correct, since one cannot identify missing functionality, such as missing nodes or edges, using these techniques. The graphical representation of the software can be accomplished at varying levels of detail to focus attention on components of the system that are in need of more (or less) intensive testing. (For more information on model-based testing, see Rosaria and Robinson, 2000.)

Test oracles,[2] in this context, are separate programs that take small sections of the sequence of user-supplied inputs and provide output that represents the proper functioning of a software system as defined by the system's requirements. (In other contexts, other test oracle architectures and configurations are possible, e.g., architectures where individual oracles cooperate to constitute an oracle for the whole. See Oshana, 1999, for details.) When they exist, test oracles are placed at each node of the graph to permit comparison of the functioning of the current system with what the output of the correct system would be. Through use of an oracle, discrepancies between the current state of the system and the correct system at various nodes of the graph can be identified.[3]

Graphical models are initially developed at a very early stage in system development, during the identification of system requirements. A number of tools (Rational, Visio) and language constructs (Unified Modeling Language) are available for creating graphical models, which, once created, can be modified based on issues raised during system development. One can also develop a procedure for the generation of test cases based on the model and then update the procedure as the system and the graphical model mature. Many users have found that the graphical model is a useful form of documentation of the system and provides a useful summary of its features.

An illustrative example of a graphical model for a long-distance telephone billing system is provided in Figure 3-1, which has two parts: (a) a model of the general process and (b) a detailed submodel of the primary component. This illustrative example suggests the hierarchical nature of these models in practice.

---

[2]Software cost reduction (SCR) can be used to produce test oracles.
[3]Since model-based testing is not a direct examination of the software code, it is considered a type of "black box" testing.

TESTING METHODS AND RELATED ISSUES 23

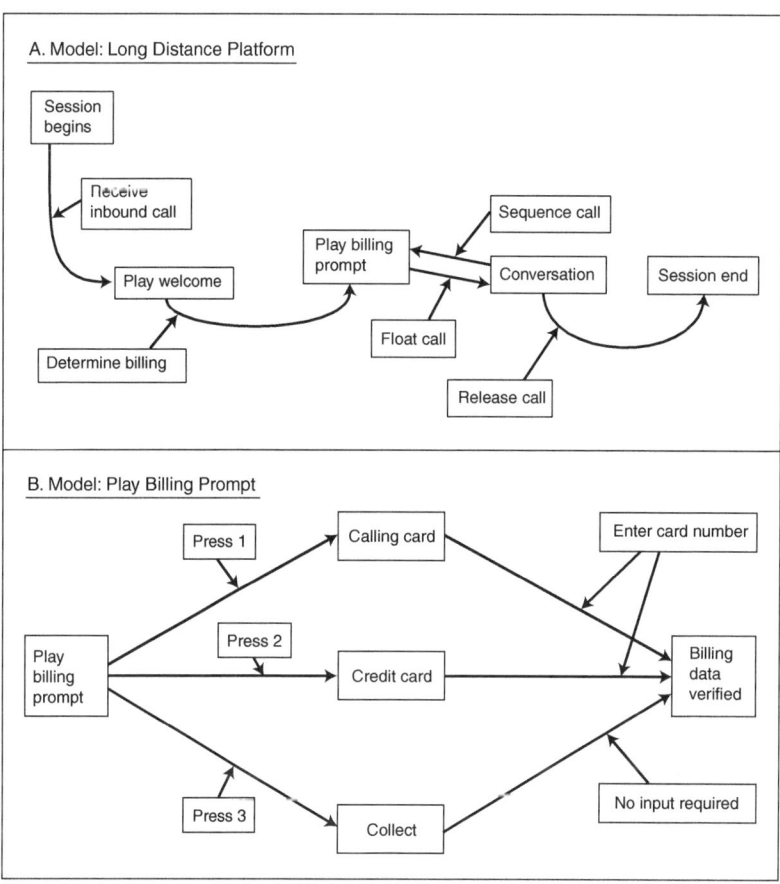

FIGURE 3-1 Example of a model used in model-based testing of long-distance telephone flows: (a) high-level model and (b) detailed submodel of billing option prompts.
SOURCE: Adapted from Apfelbaum and Doyle (1997).

Major benefits of model-based testing are early and efficient defect detection, automatic generation (very often) of test suites, and flexibility and coverage of test suites. Another primary benefit of model-based testing is that when changes are made to the system, typically only minor changes are needed to the model and thus test scenarios relative to the new system can be generated quickly. In addition, since the graphical model can be modified along with the software system, model-based testing works smoothly in concert with spiral software development. On the downside, testers need to develop additional skills, and development of the model represents a substantial up-front investment.

Testing, which requires substantial time and resources, is very costly, both for the service test agencies and, much more importantly, for the software developers themselves, and so it is essential that it be done efficiently. DoD software systems in particular are generally required to be highly dependable and always available, which is another reason that testing must be highly effective in this area of application. The testing procedure often used for defense systems is manual testing with custom-designed test cases. DoD also contracts for complex large, custom-built systems and demands high reliability of their software under severe cost pressures. Cost pressures, short release cycles, and manual test generation have the potential for negatively affecting system reliability in the field.

A recent dynamic in defense acquisition is the greater use of evolutionary procurement or spiral development of software programs. Particularly with these types of procurement, it is extremely efficient for testing to be integrated into the development process so that one has a working (sub)system throughout the various stages of system development and so that one can adjust the model to specifically test those components added at each stage.

Model-based software testing has the potential for assuring clients that software will function properly in the field and can be used for verification prior to release. The workshop presented two methods among many for model-based testing that have been shown to be effective in a wide variety of industrial applications. The first example is Markov chain usage modeling, and the second is automatic efficient test generation (AETG), often referred to as combinatorial design.[4]

MARKOV CHAIN USAGE MODELS

Markov chain usage modeling was described at the workshop by Jesse Poore of the University of Tennessee (for more details, see Whittaker and Poore, 1993, and Whittaker and Thomason, 1994). Markov chain usage models begin from the graphical model of a software program described

---

[4]We note that SCR can also be extended to automatic test set generation, albeit in a very different form than discussed here. A lot of techniques used in protocol testing also use model-based finite state machines (Lee and Yanakakis, 1992).

above. On top of this graphical model, a Markov chain probabilistic structure is associated with various user-supplied actions—shown as arcs in the graphical model—that result in transitions from one node to the nodes that are linked to it. Given the Markovian assumption, the probabilities attached to the various transitions from node to node are assumed to be (1) independent of the path taken to arrive at the given node and (2) unchanging in time. These (conditional) probabilities indicate which transitions from a given node are more or less likely based on the actions of a given type of user. Importantly, these transition probabilities can be used in a simulation to select subsequent arcs, proceeding from one node to another, resulting in a path through the graphical model.

Testing Process

The basics of the Markov chain usage model testing process are as follows. There is a population of possible paths from the initial state to the termination state(s) of a program. The Markov chain usage model randomly samples paths from this population using the transition probabilities, which are typically obtained in one of three ways: (1) elicited from experts, (2) based on field data recorded by instrumented systems, or (3) resolved from a system of constraints. By selecting the test inputs in any of these three ways, the paths that are more frequently utilized by a user are chosen for testing with higher probability. Defects associated with the more frequent-use scenarios are thus more likely to be discovered and eliminated. An important benefit of this testing is that, based on the well-understood properties of Markov chains, various long-run characteristics of system performance can be estimated, such as the reliability remaining after testing is concluded. Additional metrics based on Markov chain theory are also produced. (See Poore and Trammell, 1999, for additional details on Markov chain model-based usage testing.) A representation of a Markov chain model is shown in Figure 3-2.

While the assumptions of conditional independence and time homogeneity are capable of validation, it is not crucial that these assumptions obtain precisely for this methodology to be useful. It is very possible that the conditional independence assumption may not hold; for instance, it may be that the graphical model is at such a fine level of detail that movement from prior nodes may provide some information about movement to subsequent nodes. Also, it is possible that the time homogeneity assumption may not hold; for example knowledge of the number of nodes visited

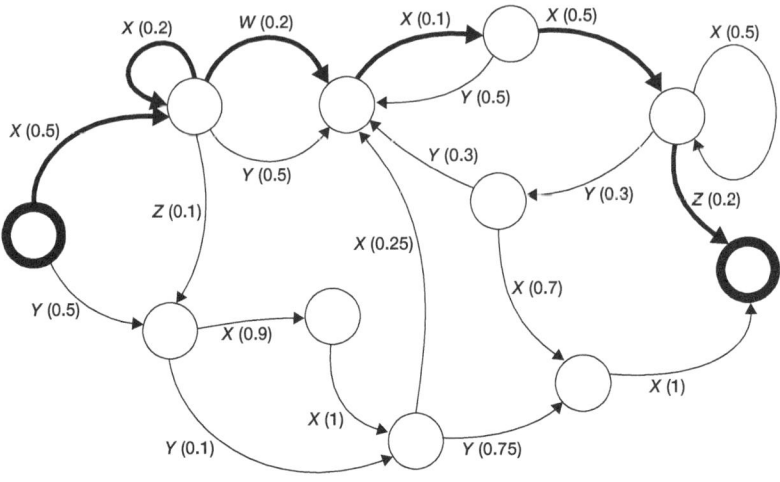

FIGURE 3-2 Example of Markov chain usage model.
SOURCE: Workshop presentation by Stacy Prowell.

prior to the current node may increase the probability of subsequent termination. However, these two assumptions can often be made approximately true by adjusting the graphical model in some way. Furthermore, even if the time homogeneity and conditional independence assumptions are violated to a modest extent, it is quite likely that the resulting set of test inputs will still be useful to support acquisition decisions (e.g., whether to release the software).

The elements of the transition probability matrix can be validated based on a comparison of the steady-state properties of the assumed Markov chain and the steady-state properties associated with anticipated use. Test plans can also be selected to satisfy various user-specified probabilistic goals, e.g., testing until each node has been "visited" with a minimum probability. The Markov chain model can also be used to estimate the expected cost and time to test. If the demand on financial resources or time is too high, the software system under test or its requirements may be modified or restructured so that testing is less expensive. Finally, the inputs can be stratified to give a high probability of selecting those inputs and paths associated with functionality where there is high risk to the user.

With defense systems, there are often catastrophic failure modes that have to be addressed. If the product developers know the potential causes of catastrophic failures and their relationship to either states or arcs of the system, then those states or arcs can be chosen to be tested with certainty. Indeed, as a model of use, one could construct the "death state," model the ways to reach it, and then easily compute statistics related to that state and the paths to it. (Markov chain usage model testing is often preceded by arc coverage testing.)

Automated test execution, assuming that the software has been well constructed, is often straightforward to apply. To carry out testing, the arcs corresponding to the test paths, as represented in the graphical model, must be set up to receive input test data. In addition, the tester must be able both to control inputs into the system under test and to observe outputs from the system. Analysis of the test results then requires the following: (1) the tester must be able to decide whether the system passed or failed on each test input, (2) since it is not always possible to observe all system outputs, unobserved outputs must be accounted for in some way, and (3) the test oracle needs to be able to determine whether the system passed or failed in each case. These requisites are by no means straightforward to obtain and often require a substantial amount of time to develop.

Stacy Prowell (University of Tennessee) described some of the tools that have been developed to implement Markov chain usage models. The Model Language (TML) supports definitions of models and related information, and allows the user to develop hierarchical modeling, with subroutines linked together as components of larger software systems; such linking is useful as it supports reuse of component software. In addition, JUMBL (Java Usage Model Building Library) contains a number of tools for analysis of Markov chain usage models, e.g., tools that compute statistics that describe the functioning of a Markov chain, such as average test length. This analysis is particularly useful for validating the model. JUMBL also contains a number of tools that generate test cases in support of various test objectives. Examples include the generation of tests constrained to exercise every arc in the model, tests generated by rank order probability of occurrence until a target total probability mass is reached, and custom-designed tests that meet contractual requirements.

An interesting alternative use of Markov chain modeling by Avritzer and Weyuker (1995) bases the transition probabilities on data collection. Another feature in Avritzer and Weyuker is the deterministic, rather than probabilistic, selection of the test suite, i.e., choosing those inputs of highest probability, thus ensuring that the most frequently traversed paths are tested.

Industrial Example of the Benefits of Markov Chain Usage Models

Markov chain usage models have been shown to provide important benefits in a variety of industrial applications. Users that have experienced success include IBM, Microsoft, U.S. Army Tank-Automative and Armaments Command, computerized thermal imaging/positron emission tomography systems, the Federal Aviation Administration Tech Center, Alcatel, Ericsson, Nortel, and Raytheon. Ron Manning reported at the workshop on Raytheon's successful application of Markov chain usage testing.

Raytheon's software development group, which has broad experience in developing large software systems, until recently used structured analysis/structured design methodology. For testing, Raytheon used unit testing, program testing, and subsystem testing. The unit testing emphasized code coverage, and the program and subsystem testing used formal test procedures based on requirements. Using this conventional approach to testing, each software system eventually worked, but many defects were found during software and system integration (even after 100 percent code coverage at the unit level). The effectiveness of unit testing was marginal, with many defects discovered at higher levels of software integration. Regression testing was attempted using the unit testing but was too expensive to complete. It was typical for approximately 6 defects per 1,000 lines of code to escape detection.

The lack of success with conventional testing was a concern for Raytheon, and the cost and time demands seemed excessive to support a successful conventional testing system. Furthermore, a great deal of effort was required for integration at each system level because of the undiscovered defects. To address these problems, cleanroom software methods, including Markov chain usage–based testing, were examined for possible use. The results of applying Markov chain usage–based testing to eight major software development projects were a greater than tenfold reduction in defects per 1,000 lines of code, software development costs within budget, and expedited system integration. For Raytheon's systems, it was found that automated test oracles could be developed to facilitate automated testing. The additional tools required were minimal, but careful staff training was found to be vital for the success of this switch in testing regimen. When this phase of the testing was completed, there was still a role for some conventional testing, but such testing was significantly expedited by the reduction in the defects in the software.

In its implementation of Markov chain usage–based testing, Raytheon's approaches included three profiles: (1) normal usage, (2) a usage model focused on error-prone modules, and (3) a usage model that explored hardware errors. It also seemed evident that the graphical models must be kept relatively simple, since the potential for growth of the state space could thus be kept under control. In this specific area of application, the typical graphical model had 50 or fewer states. Finally, while development of an automated test oracle did represent a major expense, it could be justified by the reductions in time and costs of development that resulted from the implementation of this methodology.

In addition to Markov chain usage–based testing, Raytheon initially augmented its testing with purposive sets of test inputs to cover all nodes and arcs of the graphical model for each system under test. This method was also used to debug the test oracle. It was discovered that this purposive testing was less necessary than might have been guessed, since relatively small numbers of Markov chain–chosen paths achieved relatively high path coverage. In the future, Raytheon will examine implementing usage-based testing methods at higher levels of integration. JUMBL has tool support for composing graphical models at higher levels of integration from submodels, which will correspond to the integration of systems from components. A poll of Raytheon software developers unanimously endorsed use of Markov chain–based usage testing for their next software development project.

AETG TESTING

AETG (see, e.g., Cohen et al., 1994, 1996; Dalal et al., 1998) is a combinatorial design-based approach to the identification of inputs for software testing. Consider an individual node of the graphical representation of a software system described above. A node could be a graphics-user interface in which the user is asked to supply several inputs to support some action of the software system. Upon completion, very typically based on the inputs, the software will then move to another node of the model of the software system. For this example, the user may be asked to supply categorical information for, say, seven fields. If all combinations are feasible, the possible number of separate collections of inputs for the seven fields would be a product of the seven integers representing the number of possible values for each of the fields. For even relatively small numbers of fields and values per field, this type of calculation can result in a large number of possible inputs that one might wish to test. For example, for seven dichotomous input fields, there are potentially 128 (2^7) test cases. With 13 input fields, with three choices per field, there are 1.6 million possible test cases. For most real applications, the number of fields can be much larger, with variable numbers of inputs per field.

Cohen et al. (1994) provide an example of the provisioning of a telecommunications system where a particular set of inputs consisted of 74 fields, each with many possible values, which resulted in many billions of possible test cases. Furthermore, there are often restricted choices due to constraints for inputs based on other input values, which further complicates the selection of test cases. For example, for a credit card–based transaction one needs to supply the appropriate credit card category (e.g., Mastercard, Visa, etc.) and a valid card number, while for a cash transaction those inputs have to be null. This complicates the selection of test scenarios. This issue is particularly critical since almost two-thirds of code is typically related to constraints in stopping invalid inputs. For this reason, test cases, besides testing valid inputs, also need to test invalid inputs.

As argued at the workshop by Ashish Jain of Telcordia Technologies, rather than test all combinations of valid and invalid inputs, which would often be prohibitively time-consuming, AETG instead identifies a small set of test inputs that has the following property: for each given combination of valid values for any k of the input fields, there is at least one input in the test set that includes this combination of values. In practice, k is often as small as 2 or 3. For example, in the pairwise case, for input fields I and J, there will exist in the set of test inputs at least one input that has value i for field I and value j for field J, for every possible combination of i and j and every possible combination of two fields I and J. For invalid values, a more complicated strategy is utilized.

A key empirical finding underlying this methodology is that in many applications it is the case that the large majority of software errors are expressed through the simultaneous use of only two or three input values. For example, a detailed root cause analysis of field trouble reports for a large Telcordia Technologies operation support system demonstrated that most field defects were caused by pairwise interactions of input fields. Nine system-tested input screens of a Telcordia inventory system were retested using AETG-generated test cases, and 49 new defects were discovered through use of an all-pairwise input fields test plan. Given this empirical evidence, users typically set k equal to 2 or 3.

To demonstrate the gains that are theoretically possible, for the situation of 126 dichotomous fields (a total of approximately 10^{38} paths), AETG identified a set of only 10 test cases that included all pairwise sets of inputs. More generally, with k fields, each with l possible values, AETG finds a set of inputs that has approximately $l^2 \log(k)$ members. The set of inputs identified by the AETG method has been shown, in practice, to have good code coverage properties.

Industrial Examples of the Benefits of AETG Testing

Two applications of AETG were described at the workshop. The first, presented by Manish Rathi of Telcordia Technologies, was its application to the testing of an airplane executing an aileron roll (which could be input into either an operational test or the test of an airplane simulator). The objectives of the test were to: (1) assess the pilot's ability to respond to various disturbances produced by the aileron roll, (2) detect unwanted sideslip excursions and the pilot's ability to compensate, and (3) test inertial coupling. Key inputs that affect the "rolling" characteristics of an airplane are the air speed, the Mach number, the altitude, and the position of the flaps and landing gear. (Various additional factors such as air temperature also affect the rolling characteristics but they were purposely ignored in this analysis.) For each of these four inputs, a finite number of possible values were identified for testing purposes. (All possible combinations testing was not feasible for two reasons: first, it would represent too large a set of test events to carry out; second, there were additional constraints on the inputs, prohibiting use of some combinations of input values.) Even given the constraints, there were more than 2,000 possible legal combinations of test values, i.e., more than 2,000 possible test flights. AETG was therefore used to identify a small number of test events that included inputs containing all pairwise combinations of input values for pairs of input fields (while observing the various constraints). AETG discovered 70 test flights that included tests with all possible combinations of pairwise input values for pairs of input fields, a reduction of more than 96 percent, which is extremely important given the cost of a single test flight.

The second application, described at the workshop by Jerry Huller of Raytheon, was for the software used to guide a Raytheon satellite control center combined with a telemetry, command, and ranging site, which are both used to communicate with orbiting satellites. The system contains a great deal of redundancy to enhance its overall reliability. A primary test problem is that, given the redundancy, there are many combinations of equipment units that might be used along possible signal paths from the ground system operator to the satellite and back, and it is necessary to demonstrate that typical satellite operations can be performed with any of the possible paths using various combinations of equipment units. Exhaustive testing of all combinations of signal paths was not practical in a commercial setting. An efficient way of generating a small number of test cases was needed that provided good coverage of the many possible signal paths. AETG generated test cases that covered all pairwise combinations of test inputs, and it also handled restrictions on allowable input combinations. In this application, there were 144 potential test cases of which AETG identified 12 for testing, representing a 92 percent reduction in testing. Taking into consideration some additional complications not mentioned here, the AETG strategy provided an overall 68 percent savings in test duration and a 67 percent savings in test labor costs.

INTEGRATION OF AETG AND MARKOV CHAIN USAGE MODELS

For a graphical model with limited choices at each node and for a relatively finite number of nodes, Markov chain usage model testing is a methodology that provides a rich set of information to the software tester along with an efficient method for selecting test inputs. However, the derivation of test oracles and user profiles can be complicated, especially for graphical models that have a large number of nodes and for nodes that have a large number of arcs due to many input fields and values per field. One possibility for these graphical models is to substitute, for the set of all possible transitions, just those transitions that are selected by AETG. This would reduce the number of separate nodes and arcs needed. Therefore, AETG would be used to reduce the number of paths through a software system and the Markov chain usage model would provide a probabilistic structure only for the AETG selections. Another way to combine these methods is where the usage models would handle the transitions from one node to another and AETG would determine the possible choices for user inputs at each node as they are encountered in the graphical model. In other words, AETG could operate either at the level of the entire graphical model or at the level of individual nodes. Other approaches may also be possible. This is an area in which further research would likely provide substantial benefits.

TEST AUTOMATION

Clearly, without test automation, any testing methodology, including model-based testing methods, will be of limited use. All of the steady-state validation benefits and estimates of the costs and number of replications needed for Markov chain usage model testing are predicated on the ability to carry out a relatively large number of tests. Reliability projections that can be computed based on usage models reveal that the number of tests needed to demonstrate reliability of 99 percent or more are almost always beyond the budget for most complicated systems. Therefore, since test automation is a virtual necessity for modern software testing, the topic was examined at the workshop.

Mike Houghtaling of IBM presented a number of automation tools that the company has used in managing its tape systems and libraries. The company uses Markov chain usage models comprising roughly 2,000 states, which is difficult to represent on paper. A software tool called TeamWork is therefore used to provide a graphical capability for constructing the collection of state transition diagrams. There is also a tool for the implementation of the test oracles. For selection of test cases through random sampling and for composing the test results and generating the summary measures, ToolSet_Certify is used. ToolSet_SRE is used to break up the model into subsystems for focused sampling and testing and to reaggregate the subsystems for full system analysis. To manipulate the device drivers for the test execution, CBasc is used. In addition, CORBA middleware (in particular ORBLink) is used to increase automation, and QuickSilver is used as a test visualizer to help depict results in a graphical environment.

Houghtaling noted several complicating factors with automation. First, three concurrent oracle process strategies need to be supported: (1) postoracle analysis, (2) concurrent oracle analysis, and (3) preoracle certificate generation. Second, several similar activities need to be separately addressed: (1) failure identification versus defect diagnosis, since the cause of the failure may have happened many events prior to the evident failure; (2) usage and model abstractions versus implementation knowledge, because implementers sometimes make assumptions about uses or failures that are inconsistent with new uses of systems (for example, some years ago data began to flow over telephone systems); and (3) test automation component design for interoperability versus analysis of the functioning of the complete system. (Some test automation equipment is designed primarily to watch the interface between two components so that errors are not made in the interface itself, but there may not be automated support for testing the full system.) Besides these, there is an obvious challenge in restoring the environment and preconditions before the tests can be run.

The separation of failure identification and defect diagnosis is important since fault diagnosis requires in-depth knowledge of the system implementation. Test teams, especially black box–oriented and system-level test teams, might not possess this knowledge. Test automation tools should also be composed from the same perspective. Tools that are directed toward usage certifications and that are consequently failure profile–oriented should not be encumbered with diagnostic-oriented artifacts. The ability to leverage the information collected during failure identification, such as automatically replaying the usage scenario in an execution environment that is configured with diagnostic tools, is beneficial, but should not subvert the usage testing environment.

In addition, designers and implementers of test automation tools employing high-level usage models need to be aware of the gap between the usage model vocabulary and the more concrete implementation vocabulary that is meaningful to the developers of the system. Attempts to document justifications for claims about test case failures will need to bridge the gap.

Finally, with respect to test automation component design for interoperability, a cost-effective test automation environment will need to be based on industry standards and possess the capability of collaborating within a framework supporting many of the aspects associated with development and test processes (planning, domain modeling, test selections, test executions and evaluations, system assessments, and test progress management).

METHODS FOR TESTING INTEROPERABILITY

It is typical for software-intensive systems to comprise a number of separate software components that are used interactively, known as a system of systems. Some of these component systems may be commercial-off-the-shelf (COTS) systems, while others may be developed in-house. Each of these software components is subject to separate update schedules, and each time a component system is modified there is an opportunity for the overall software system to fail due to difficulties the modified component system may have in interacting with the remaining components. Difficulties in interactions between components are referred to as interoperability failures. (The general problem involves interaction with hardware as well as software components, but the focus here is on software interactions.) A session at the workshop addressed this critical issue given its importance in Department of Defense applications; the presentations examined general tools to avoid interoperability problems in system development, and how one might test the resulting system of systems for defects.

Amjad Umar of Telcordia Technologies discussed the general interoperability problem and tools for its solution, which he referred to as integration technologies. Consider as an example an e-business activity made up of several component applications. Customer purchases from this e-business involve anywhere from a dozen to hundreds of applications, which need to interoperate smoothly to complete each transaction. Such interoperability is a challenge because transactions may involve some systems that are very new as well as some that are more than 20 years old. In addition to being of different vintages, these components may come from different suppliers and may have either poor or nonexistent documentation. Integration is typically needed at several levels, and if the systems are not well integrated, the result can be an increase in transaction errors and service time. On the other hand, if the systems are well integrated, human efforts in completing the transaction will be minimized. The overall challenge then is to determine good procedures for integrating the entire system, to set the requirements to test against, and to test the resulting system.

A considerable amount of jargon is associated with this area. A glossary is included in Appendix B that defines some of the more common terms used in this context.

Amjad Umar's presentation provided a general framework for the wide variety of approaches and techniques that address interoperability problems. This is possible because the number of distinct concepts in this area is much more finite than the jumble of acronyms might suggest. The overall idea is to develop an "integration bus" with various adapters so that different applications can plug into this bus to create a smoothly interactive system.

Integration Technologies

To start, integration is needed at the following levels: (1) cross-enterprise applications, (2) internal process management, (3) information transformation, (4) application connectivity, and (5) network connectivity. Solution technologies that can be implemented at these various layers include, from high to low: (1) business-to-business integration platforms, (2) enterprise application integration platforms, (3) converters, (4) middleware and adapters, and (5) network transport.

With respect to low-level integration between two systems, interconnection technologies might work as a mediator between the order processing, provided by a Web browser or Java applet, and the inventory, managed by user interfaces, application code, and a data source. Interconnection technologies may involve a remote user connector, a remote method connector, and a remote data connector. At midlevel integration, there are object wrappers (e.g., CORBA) that function in combination with screen scrapers, function gateways, and data gateways (e.g., ODBC). At high-level integration, but within an enterprise, software tools such as adapters provide a smooth integration between order processing and the inventory system. Finally, at very high-level integration, again tools such as several kinds of adapters smooth the integration between inventory systems and order processing systems and trading hubs across firewalls for external organizations.

General platforms that attempt to minimize integration problems are available to oversee the process; leading examples include Sun's J2EE platform, IBM's e-business framework, and Microsoft's .NET. The choice of integration technology to implement depends on the number of applications, the flexibility of requirements, the accessibility of applications, and the degree of organizational control.

Clearly, integration raises persistent challenges. It is popular because it permits use of existing software tools. On the other hand, it is problematic because it increases the workload on existing applications, may not be good for long-range needs, and creates difficult testing challenges (however, see below for a possible approach to testing systems of systems). Many tools are being developed to address these problems, but they are specific to certain areas of application, and whether they will be portable to the defense environment is an important question to address. Also, it is important to note that addressing interoperability problems may sometimes not be cost-effective compared to scrapping an old system and designing all system components from scratch.

Additional information on integration technologies can be found in Lithicum (2001) and Umar (2002, 2003). Information can also be found at the following Web sites: (1) www.vitria.com, (2) www.webmethods.com, and (3) www.tibco.com.

Testing a System of Systems for Interoperability

Consider a system that is a mixture of legacy and new software components and that addresses interoperability through use of middleware technology (e.g., MOM, CORBA, or XML). Assume further that the application is delivered using Web services, and so is composed of services from components across the Internet or other secured networks. As an example, consider a survivable, mobile, ad hoc tactical network with a variety of clients, namely infantry and manned vehicles, that need to obtain information from a command and information center, which is composed of various types of servers. These are often high-volume and long-distance information transactions. The distributed services are based on a COTS system of systems with multiple providers. These systems themselves evolve over time, producing a substantial amount of code and functionality churn. The corresponding system can thus get very complex.

In an example described at the workshop by Siddhartha Dalal, for a procurement service created on a system consisting of many component systems, there was a procurer side, a provider side, and a gateway enabling more than 300,000 transactions per month with 10,000 constantly changing rules governing the transactions. There were additions and deletions of roughly 30 rules a day, and there was on average one new software component release per day. As a result of these dynamics, 4,000 errors in processing occurred in 6 months, but, using traditional testing methods, only 47 percent of the defects ultimately discovered were removed during testing.

The problem does not have an easy solution. In interviews with a number of chief information officers responsible for Web-based services, the officers stated that service failures reported by customers had the following distribution of causes:

(1) lack of availability (64 percent)
(2) software bugs (55 percent)
(3) bad or invalid data (47 percent)
(4) software upgrade failure (46 percent)
(5) incorrect, unapproved, or illegal content (15 percent)
(6) other (5 percent)

(Some of these causes can occur simultaneously, thus the sum of the percentages is greater than 100 percent.)

In moving from software products to component-based services, the traditional preproduction product-based testing fails because it is not focused on the most typical sources of error from a service perspective. When testing a system of systems, it is critical to understand that traditional testing does not work for two reasons: first, the owner of a system of systems does not have control over all the component systems; second, the component systems or the rules may change from one day to the next.

While product testing usually only needs to implement design for testability with predictable churn, in the application of Web-based services one needs a completely different testing procedure designed for continuous monitorability, with a service assurance focus, to account for *un*predictable churn.

With this new situation, there are usually service-level agreements concentrating on availability and performance based on end-to-end transactions. To enforce the service agreements, one needs constant monitoring of the deployed system (i.e., postproduction monitoring) with test transactions of various types. However, one cannot send too many transactions to test and monitor the system as they may degrade the performance of the system.

One approach for postproduction monitoring and testing was proposed by Siddhartha Dalal of Telcordia Technologies (see Dalal et al., 2002). In this approach, a number of end-to-end transactions are initially captured, and then a number of synthetic end-to-end transactions are generated. The idea is to generate a minimal number of user-level synthetic transactions that are very sensitive for detecting functional errors and that provide 100 percent pairwise functional coverage at the node level. This generation of synthetic end-to-end transactions can be accomplished through use of AETG.

To see this, consider an XML document with 50 tags.[5] Assuming only two values per tag, this could produce 2^{50} possible test cases. The AutoVigilance system, which was created by Dalal and his colleagues at Telcordia, produces nine test cases from AETG that cover all pairwise choices of tags in this case. These synthetic transactions are then sent through probes, with the results analyzed automatically and problems proactively reported using alerts. Finally, possible subsystems causing problems are identified using an automated root cause analysis. This entire

---

[5]A tag in XML identifies and delimits a text field.

process can be automated and is extremely efficient for nonstop monitoring and testing as well as regression testing when needed. The hope, by implementing this testing, is to find problems before end-users do.

Based on the various methods discussed during this session, the panel concluded that model-based testing offers important potential for improving the software testing methods currently utilized by the service test agencies. Two specific methods, Markov chain usage–based testing and AETG, were described in detail, and their utility in defense or defense-type applications was exhibited. There are other related approaches, some of which were also briefly mentioned above. The specific advantages and disadvantages from widespread implementation of these relatively new methods for defense software systems needs to be determined. Therefore, demonstration projects should be carried out.

Test automation is extremely important to support the broad utility of model-based methods, and a description of automation tools for one specific application suggests that such tools can be developed for other areas of application.

Finally, interoperability problems are becoming a serious hurdle for defense software systems that are structured as systems of systems. These problems have been addressed in industry, and various tools are now in use to help overcome them. In addition, an application of AETG was shown to be useful in discovering interoperability problems.

4

Data Analysis to Assess Performance and to Support Software Improvement

The model-based testing schemes described above will produce a collection of inputs to and outputs from a software system, the inputs representing user stimuli and the outputs measures of the functioning of the software. Data can be collected on a system either in development or in use, and can then be analyzed to examine a number of important aspects of software development and performance. It is important to use these data to improve software engineering processes, to discover faults as early as possible in system development, and to monitor system performance when fielded. The main aspects of software development and performance examined at the workshop include: (1) measurement of software risk, (2) measurement of software aging, (3) defect classification and analysis, and (4) use of Bayesian modeling for prediction of the costs of software development. These analyses by no means represent all the uses of test and performance data for a software system, but they do provide an indication of the breadth of studies that can be carried out.

MEASURING SOFTWARE RISK

When acquiring a new software system, or comparing competing software systems for acquisition, it is important to be able to estimate the risk of software failure. In addressing risk, one assumes that associated with each input i to the software system there is a cost resulting from the failure of the software. To determine which inputs will and will not result in system failure, a set of test inputs is typically selected with a contractual understanding to complete the entire test set (using some test selection method) and the software is then run using that set. If, for various reasons, a test regimen ends up incomplete, this incompleteness needs to be accounted for to provide an assessment of the risk of failure for the software. The interaction of the test selection method, the sometimes incomplete process of testing for defects, the probability of selection of inputs by users, and the fact that certain software failures are more costly than others all raise some interesting measurement issues, which were addressed at the workshop by Elaine Weyuker of AT&T.

To begin, one can assume either that there is a positive cost of failure, denoted $cost(i)$ associated with every input i, or that there is a cost $c(i)$ that is positive only if that input actually results in a system failure, with the cost $c(i)$ being set equal to zero otherwise. (In other words, $cost(i)$ measures the potential consequences of various types of system failure, regardless of whether the system would actually fail, and $c(i)$ is a realized cost that is zero if the system works with input i.) A further assumption is that one can estimate the probability that various test inputs occur in field use; such inputs are referred to collectively as the operational distribution.

Assume also that a test case selection method has been chosen to estimate system risk. This can be done, as discussed below, in a number of different ways. The selection of a test suite can be thought of as a contractual agreement that each input contained in the suite must be tried out on the software. A reasonable and conservative assumption is that any test cases not run are assumed to have failed had they been applied. This assumption is adopted to counter the worry that one could bias a risk assessment by not running cases that one thought a priori might fail. Any other missing information is assumed to be replaced by the most conservative possible value to provide an upper bound on the estimation of risk. In this way, any cases that should have been run, but weren't, are accounted for either as if they have been run and failed or as if the worst possible case has been run and failed, depending on the contract. The cost of running a software program on test case i is therefore defined to be $c(i)$ if the program is run on i, and is defined to be $cost(i)$ otherwise, using this conservative assumption. The overall estimated risk associated with a software program, based on testing using some test suite, is defined to be the weighted sum over test cases of the product of the cost of failure ($c(i)$ or $cost(i)$) for test input i multiplied by the (normalized) probability that test input i would occur in the field (normalized over those inputs actually contained in the test suite when appropriate, given the contract).

Obviously, it is very difficult or impossible to associate a cost of failure with every possible input, since the input domain is almost always much too large. Besides, even though the test suite is generally much smaller than the entire input domain, it can be large, and as a result associating a cost with every element of the test suite can be overwhelming. However, once the test suite has been run, and one can observe which inputs resulted in failure, one is left with the job of determining the cost of failure for only a very small number of inputs, those that have been run and failed. This is an important advantage of this approach. One also knows how the system failed and therefore the assignment of a cost should be much more feasible.

Weyuker outlined several methods that could be used to select inputs for testing, categorized into two broad categories: (1) statistical testing, where the test cases are selected (without replacement) according to an (assumed) operational distribution, and (2) deterministic testing, where purposively selected cases represent a given fraction of the probability of all test cases, sorted either by probability of use or by the risk of use (which is the product of the *cost(i)* and the probability of use), with the highest p percent of cases selected for testing, for some p. The rationale behind this is that these are the inputs that are going to be used most often or of highest risk and if these result in failure, they are going to have a large impact on users. In this description we will focus on statistical testing, though examples for deterministic testing were also discussed.

Under statistical testing, Weyuker distinguished between (a) using the operational distribution and (b), in a situation of ignorance, using a uniform distribution over all possible inputs. She made the argument that strongly skewed distributions were the norm and that assuming a uniform distribution as a result of a lack of knowledge of the usage distribution could strongly bias the resulting risk estimates. This can be clarified using the following example, represented by the entries in Table 4-1.

Assume that there are 1,000 customers (or categories of customers) ranked by volume of business. These make up the entire input domain. Customer i_1 represents 35 percent of the business, while customer $i_{1,000}$ provides very little business. Assume a test suite of 100 cases was selected to be run using statistical testing based on the operational distribution, but only 99 cases were run (without failure); i_4, the test case with the largest risk and not selected for testing was not run. Then we behave as if i_4 is the

TABLE 4-1 Example of Costs and Operational Distribution for Fictitious Inputs

Input	Pr	C	Pr × C
i_1	0.35	5,000	1,750
i_2	0.20	4,000	800
i_3	0.10	1,000	100
i_4	0.10	100	10
i_5	0.10	50	5
i_6	0.07	40	2.8
i_7	0.03	50	1.5
i_8	0.01	100	1.0
i_9	0.005	5,000	25.0
i_{10}	0.005	10	0.05
i_{11}	0.004	10	0.04
i_{12}	0.003	10	0.03
i_{13}	0.003	1	0.003
$i_{14} - i_{100}$	0.01999	1	0.01999
$i_{101} - i_{999}$	0.00001	1	0.00001
$i_{1,000}$	10^{-7}	10^9	100

input that had been selected as the 100th test case and that it failed. The risk associated with this software is therefore 100 × 0.10 (1.00/0.9999899), or roughly 10.

If one instead (mistakenly) assumes that the inputs follow a uniform distribution, with 100 test cases contracted for, then if 99 test cases were run with no defects, the risk is 1/100 times the highest $c(i)$ for an untested input i, or in this case 10^7. Here the risk estimate is biased high since that input is extremely rare in reality.

Similar considerations arise with deterministic selection of test suites. A remaining issue is how to estimate the associated field use probabilities, especially, as is the case in many situations, where the set of possible inputs or user types is very large. This turns out to be feasible in many situations: AT&T, for example, regularly collects data on its operational distributions for large projects. In these applications, the greater challenge is to model the functioning of the software system in order to understand the risks of failure.

FAULT-TOLERANT SOFTWARE: MEASURING SOFTWARE AGING AND REJUVENATION

Highly reliable software is needed for applications where defects can be catastrophic—for example, software supporting aircraft control and nuclear systems. However, trustworthy software is also vital to support common commercial applications, such as telecommunication and banking systems. While total fault avoidance can at times be accomplished through use of good software engineering practices, it can be difficult or impossible to achieve for particularly large, complex systems. Furthermore, as discussed above, it is impossible to fully test and verify that software is fault-free. Therefore, instead of fault-free software, in some situations it might be more practical to consider development of fault-tolerant software, that is, software that can accommodate deficiencies. While hardware fault tolerance is a well-understood concept, fault tolerance is a relatively new, unexplored area for software systems. Many techniques are showing promise for use in the development of fault-tolerant software, including design diversity (parallel coding), data diversity (e.g., *n*-copy programming), and environmental diversity (proactive fault management). (See the glossary in Appendix B for definitions of these terms.)

Efforts to develop fault-tolerant software have necessitated attempts to classify defects, acknowledging that different types of defects will likely require different procedures or techniques to achieve fault tolerance. Consider a situation where one has an availability model with hardware redundancy and imperfect recovery software. Failures can be broadly classified into recovery software failures, operating system failures, and application failures. Application failures are often dealt with by passive redundancy, using cold replication to return the application to a working state. Software aging[1] occurs when defect conditions accumulate over time, leading to either performance degradation or software failure. It can be due to deterioration in the availability of operating system resources, data corruption, or numerical error accumulation. The use of design diversity to address software aging can often be prohibitively expensive. Therefore environmental diversity, which is temporal or time-related diversity, may often be the preferred approach.

---

[1]Note that use of the term "software aging" is not universal; the problem under discussion is also considered a case of cumulative failure.

One particular type of environmental diversity, software rejuvenation, which was described at the workshop by Kishor Trivedi, is restarting an application to return to an initializing state. Rejuvenation incurs some costs, such as downtime and lost transactions, and so an important research issue is to identify optimal times for rejuvenation to be carried out. There are currently two general approaches to scheduling rejuvenation: those based on analytical modeling, and those based on measurement-based (empirical, statistical) rejuvenation. In analytical modeling, transactions are assumed to arrive according to a homogeneous Poisson process; they are queued and the buffer is of finite size. Transactions are served by an assumed nonhomogeneous Poisson process (NHPP) and the software failure process is also assumed to be NHPP. Based on this model, two rejuvenation strategies that have been proposed are a time-based approach (restart the application every t_0 time periods), and a load- and time-based approach.

A measurement-based approach to scheduling rejuvenation attempts to directly detect "aging." In this model, the state of operating system resources is periodically monitored and data are collected on the attributes responsible for the performance of the system. The effect of aging on system resources is quantified by constant measurement of these attributes, typically through an estimation of the expected time to exhaustion. Again, two approaches have been suggested for use as decision rules on when to restart an application. These are time-based estimation (see, e.g., Garg et al., 1998) and workload-based estimation (see, e.g., Vaidyanathan and Trivedi, 1999). Time-based estimation is implemented by using nonparametric regressions on time of attributes such as the amount of real memory available and file table size. Workload-based estimation uses cluster analysis based on data on system workload (cpuContextSwitch, sysCall, pageIn, pageOut) in order to identify a small number of states of system performance. Transitions from one state to another and sojourn times in each state are modeled using a Markov chain model. The resulting model can be used to optimize some objective function as a function of the decision rule on when to schedule rejuvenation; one specific method that accomplishes this is the symbolic hierarchical automated reliability and performance estimator.

DEFECT CLASSIFICATION AND ANALYSIS

It is reasonable to expect that the information collected on field performance of a software system should provide useful information about both the number and the types of defects that the software contains. There are now efforts to utilize this information as part of a feedback loop to improve the software engineering process for subsequent systems. A leading approach to operating this feedback loop is referred to as orthogonal defect classification (ODC), which was described at the workshop by its developer, Ram Chillarege. ODC, created at IBM and successfully used at Motorola, Telcordia, Nortel, and Lucent, among others, utilizes the defect stream from software testing as a source of information on both the software product under development and the software engineering process. Based on this classification and analysis of defects, the overall goal is to improve not only project management, prediction, and quality control by various feedback mechanisms, but also software development processes.

Using ODC, each software defect is classified using several categories that describe different aspects of the defects (see Dalal et al., 1999, for details). One set of dimensions that has been utilized by ODC is as follows: (1) life cycle phase when the defect was detected, (2) the defect trigger, i.e., the type of test of activity (e.g., system test, function test, or review inspection) that revealed the defect, (3) the defect impact (e.g., on instability, integrity/security, performance, maintenance, standards, documentation, usability, reliability, or capability), (4) defect severity, (5) defect type, i.e., the type of software change that needed to be made, (6) defect modifier (either missing or incorrect), (7) defect source, (8) defect domain, and (9) fault origin in requirements, design, or implementation. ODC separates the development process into various periods, and then examines the nine-dimensional defect profile by period to look for significant changes. These profiles are linked to potential changes in the system development process that are likely to improve the software development process. The term orthogonal in ODC does not connote mathematical orthogonality, but simply that the more separate the dimensions used, the more useful they are for this purpose.

Vic Basili (University of Maryland), in an approach similar to that of ODC, has also examined the extent to which one can analyze the patterns of defects made in a company's software development in order to improve the development process in the future. The idea is to support a feedback loop that identifies and examines the patterns of defects, determines the leading causes of these defects, and then identifies process changes likely to reduce rates of future defects. Basili refers to this feedback loop as an "experience factory."

Basili makes distinctions among the following terms. First, there are *errors*, which are made in the human thought process. Second, there are *faults*, which are individual, concrete manifestations of the errors within the software; one error may cause several faults and different errors may cause identical faults. Third, there are *failures*, which are departures of the operational software system behavior from user expectations; a particular failure may be caused by several faults, and some faults may never result in a failure.[2]

The experience factory model is an effort to examine how a software development project is organized and carried out in order to understand the possible sources of errors, faults, and failures. Data on system performance are analyzed and synthesized to develop an experience base, which is then used to implement changes in the approach to project support.

Experience factory is oriented by two goals. The first is to build baselines of defect classes; that is, the problem areas in several software projects are identified and the number and origin of classes of defects assessed. Possible defect origins include requirements, specification, design, coding, unit testing, system testing, acceptance testing, and maintenance. In addition to origin, errors can also be classified according to algorithmic fault; for example, problems can exist with control flow, interfaces, and data definition, initialization, or use.

Once this categorization is carried out and the error distributions by error origin and algorithmic fault are well understood, the second goal is to find alternative processes that minimize the more common defects. Hypotheses concerning methods for improvement can then be evaluated through controlled experimentation. This part of the experience factory is relatively nonalgorithmic. The result might be the institution of cleanroom techniques or greater emphasis on understanding of requirements, for example.

By using experience factory models in varying areas of application, Basili has discovered that different software development environments have very distinct patterns of defects. Further, various software engineering techniques have different degrees of effectiveness in remedying various types of error. Therefore, experience factory has the potential to provide important improvements for a wide variety of software development environments.

---

[2]In the remainder of this report, the term *defect* is used synonymously with failure.

BAYESIAN INTEGRATION OF PROJECT DATA AND EXPERT JUDGMENT IN PARAMETRIC SOFTWARE COST ESTIMATION MODELS

Another use of system performance data is to construct parametric models to estimate the cost and time to develop upgrades and new software systems for related applications. These models are used for system scoping, contracting, acquisition management, and system evolution. Several cost-schedule models are now widely used, e.g., Knowledge Plan, Price S, SEER, SLIM, COCOMO II: COSTAR, Cost Xpert, Estimate Professional, and USC COCOMO II.2000. The data used to support such analyses include the size of the project, which is measured in anticipated needed lines of code or function points, effort multipliers, and scale factors.

Unfortunately, there are substantial problems in the collection of data that support these models. These problems include disincentives to provide the data, inconsistent data definitions, weak dispersion and correlation effects, missing data, and missing information on the context underlying the data.

Data collection and analysis are further complicated by process and product dynamism, in particular the receding horizon for product utility, and software practices that do not remain static over time. (For example, greater use of evolutionary acquisition complicates the modeling approach used in COCOMO II.) As a system proceeds in stages from a component-based system to a commercial-off-the-shelf system to a rapid application development system to a system of systems, the estimation error of these types of models typically reduces as a system moves within a stage but typically increases in moving from one stage to another.

If these problems can be overcome, Barry Boehm (University of Southern California [USC]), reporting on joint work with Bert Steece, Sunita Chulani, and Jongmoon Baik, demonstrated how parametric software estimation models can be used to estimate software costs. The steps in the USC Center for Software Engineering modeling methodology are: (1) analyze the existing literature, (2) perform behavioral analyses, (3) identify the relative significance of various factors, (4) perform expert-judgment Delphi assessment and formulate an a priori model, (5) gather project data, (6) determine a Bayesian a posteriori model, and (7) gather more data and refine the model. COCOMO II demonstrates that Bayesian models can be effectively used, in a regression framework, to update expert opinion using data from the costs to develop related systems. Using this methodology,

COCOMO II has provided predictions that are typically within 30 percent of the actual time and cost needed.

COCOMO II models the dependent variable, which is the logarithm of effort, using a multiple linear regression model. The specific form of the model is:

$$\ln(PM) = \beta_0 + \beta_1 \cdot 1.01 \cdot \ln(Size) + \beta_2 SF_1 \cdot \ln(Size) + \ldots + \beta_6 SF_5 \cdot \ln(Size) + \beta_7 \cdot \ln(EM_1) + \beta_8 \cdot \ln(EM_2) + \ldots + \beta_{22} \cdot \ln(EM_{16}) + \beta_{23} \cdot \ln(EM_{17})$$

where the *SFs* are scale factors, and the *EMs* are effort multipliers.

The following covariates are used: (1) the logarithm of effort multipliers (product, platform, people, and various project factors including complexity, database size, and required reliability levels) and (2) the logarithm of size effects (where size is defined in terms of thousand source lines of code) times process scale factors (constraints, quality of system architecture, and process maturity factors). The parameters associated with various covariates are initially derived from collective expert subjective opinion (using a Delphi system). Then a Bayesian approach is used to update these estimated parameters based on the collection of new data. The Bayesian approach provides predictions that approximate collective expert opinions when the experts agree, and that approximate the regression estimates when the data are clean and the experts disagree.

The specific form of the Bayesian approach is:

$$\hat{\beta} = \left[\frac{1}{s^2} X^T X + H \right]^{-1} \left[\frac{1}{s^2} X^T X + H \beta_0 \right]$$

where X represents the covariates, H is the variance of the prior for the regression coefficients, s^2 is the residual variance, and β_0 represents the prior mean.

The Bayesian updating has been shown to provide increased accuracy over the multiple regression approach. (For further information, see Baik, Boehm, and Steece, 2002; Boehm et al., 2000; and Chulani et al., 2000). New plans for COCOMO II include possible inclusion of ideas from orthogonal defect classification (see below) and from experience factory-type analyses.

In conclusion, there is a great variety of important analyses that can exploit information collected on software performance. The workshop demonstrated four applications of such data: estimation of software risk, estimation of the parameters of a fault-tolerant software procedure, creation of a feedback loop to improve the software development process, and estimation of the costs of the development of future systems. These applications provide only a brief illustration of the value of data collected on software functioning. The collection of such data in a way that facilitates its use for a wide variety of analytic purposes is obviously extremely worthwhile.

5

Next Steps

The Workshop on Statistical Methods in Software Engineering for Defense Systems, as is typical of workshops, had necessarily limited goals. Presentations reviewed a group of mainly statistically oriented techniques in software engineering that have been shown to be useful in industrial settings but that to date have not been widely adopted by the defense test and evaluation community (though there have been many initial efforts, as suggested above). The goal therefore was to make the defense test and evaluation community more aware of these techniques and their potential benefits.

As was pointed out in the introduction, it is well recognized that the software produced for software-intensive defense systems often is deficient in quality, costs too much, and is delivered late. Addressing these problems will likely require a broad effort, including, but not limited to, the greater utilization of many well-established techniques in software engineering that are widely implemented in industry. Some of the more statistically oriented techniques were described at the workshop, though the workshop should not be considered to have provided a comprehensive review of techniques in software engineering. Methods were described in the areas of: (1) development of higher-quality requirements, (2) software testing, and (3) evaluation of software development and performance.

While it seems clear that the methods described at the workshop, if utilized, would generally improve defense software development, the costs of adoption of many of the techniques need to be better understood in defense applications, as well as the magnitude of the benefits from their use, prior to widespread implementation. The panel also recognizes that Department of Defense (DoD) systems and DoD acquisition are somewhat different from typical industrial systems and development procedures; DoD systems tend to be larger and more complex, and the acquisition process involves specific milestones and stages of testing. Acceptance of any new techniques will therefore depend on their being "proven" in a slightly different environment, and on surmounting comfort with the status quo.

It also seems safe to assume, given the widespread problems with software development in the defense test and evaluation community, that this community has limited access to software engineers familiar with state-of-the-art techniques. If some of the techniques described in this report turn out to be useful in the arena of defense test and evaluation, their widespread adoption will be greatly hampered if the current staff has little familiarity with these newer methods. Therefore, the adoption of the techniques described in this report (as well as other related techniques) needs to be supported by greater expertise in-house at the service test agencies and at other similar agencies in the defense test and acquisition community. The hope is to develop either within or closely affiliated to each service test agency access to state-of-the-art software engineering expertise.

The following two recommendations address the foregoing concerns by advocating that case studies be undertaken to clearly demonstrate the costs and benefits associated with greater use of these techniques with defense systems.

Recommendation 1: Given the current lack of implementation of state-of-the-art methods in software engineering in the service test agencies, initial steps should be taken to develop access to—either in-house or in a closely affiliated relationship—state-of-the-art software engineering expertise in the operational or developmental service test agencies.

Such expertise could be acquired in part in several ways, especially including internships for doctoral students and postdoctorates at the test and evaluation agencies, and with sabbaticals for test and evaluation agency staff at industry sites where state-of-the-art techniques are developed and used.

In addition, the sessions on data analysis to assess the performance of a software system and the software development process demonstrated the broad value of analysis of data collected on both the system development process and on system performance, both from tests and from operational use. Clearly, a prerequisite to these and other valuable analyses is the collection of and facilitated access to this information in a software data archive. The service test agencies can collect information on system performance. However, collection of data on software system development—e.g., data that would support either experience factory or orthogonal defect classification, as well as data on costs, defects, various kinds of metadata—will require specification of this additional data collection in acquisition contracts. This would go well beyond collection of problem reports, in that substantial environmental information would also be collected that would support an understanding of the source of any defects.

> **Recommendation 2: Each service's operational or developmental test agency should routinely collect and archive data on software performance, including test performance data and data on field performance. The data should include fault types, fault times and frequencies, turnaround rate, use scenarios, and root cause analysis. Also, software acquisition contracts should include requirements to collect such data.**

A session at the workshop was devoted to the development of consistent, complete, and correct requirements. Two specific methods were described, which have been shown in industry and in selected defense applications to be extremely helpful in identifying errors in requirement specifications. Though only two methods were presented, there are competing methods related to those described. There is every reason to expect that widespread adoption of the methods presented or their competitors would have important benefits for developing higher-quality defense software systems. However, some training costs would be incurred with a change to these procedures. Therefore, pilot tests should be undertaken to better understand the benefits and costs of implementation.

> **Recommendation 3: Each service's operational or developmental test agency should evaluate the advantages of the use of state-of-the-art procedures to check the specification of requirements for a relatively complex defense software-intensive system.**

One effective way of carrying out this evaluation would be to develop specifications in parallel, using the service's current procedures, so that quality metric comparisons can be made. Each service would select a software system that (1) requires field configuration, (2) has the capabilities to adapt and evolve to serve future needs, and (3) has both hardware and software that come from multiple sources.

Another session at the workshop was devoted to model-based testing techniques. Clearly, both Markov chain usage testing and combinatorial design-based testing are promising for DoD applications and are likely to result in reduced software development time and higher-quality software systems. More generally, model-based testing is a general approach that could provide great advantages in testing defense software systems. There are other innovative testing strategies that are not model-based but are instead code-based, such as data flow testing (Rapps and Weyuker, 1985). To better understand the extent to which these methods can be effective, using current DoD personnel, on DoD software-intensive systems, pilot projects should be carried out, in each service, to evaluate their benefits and costs.

> **Recommendation 4: Each service's operational or developmental test agency should undertake a pilot study in which two or more testing methods, including one model-based technique, are used in parallel for several software-intensive systems throughout development to determine the amount of training required, the time needed for testing, and the method's effectiveness in identifying software defects. This study should be initiated early in system development.**

Recommendations 3 and 4 propose that each of the services select software-intensive systems as case studies to evaluate new techniques. By doing so, test and evaluation officials can judge both the costs of training engineers to use the new techniques—the "fixed costs" of widespread implementation—and the associated benefits from their use in the DoD acquisition environment. Knowledge of these costs and benefits will aid DoD in deciding whether to implement these techniques more broadly.

There are generally insufficient funds in the budgets of the service test agencies to support the actions we recommend. Also, due to their priorities, it is very unlikely that program managers will be interested in funding these pilot projects. Thus, support will be needed from the services themselves or from DoD in order to go forward. The committee suggests that DoD support these projects out of general research and development funds.

Recommendation 5: DoD should allocate general research and development funds to support pilot and demonstration projects of the sort recommended in this report in order to identify methods in software engineering that are effective for defense software systems in development.

The panel notes that constraints hinder DoD from imposing on its contractors specific state-of-the-art techniques in software engineering and development that are external to the technical considerations of the costs and benefits of the implementation of the techniques themselves. However, the restrictions do not preclude the imposition of a framework, such as the capability maturity model.

Recommendation 6: DoD needs to examine the advantages and disadvantages of the use of methods for obligating software developers under contract to DoD to use state-of-the-art methods for requirements analysis and software testing, in particular, and software engineering and development more generally.

The techniques discussed in this report are consistent with these conditions and constraints. We also note that many of the techniques described in this report are both system oriented and based on behavior, and are therefore applicable to both hardware and software components, which is an important advantage for DoD systems.

References

Aerospace Daily
   2003   Revenue for defense IT expected to grow 10-12 percent. 206(4):7.

Apfelbaum, L., and J. Doyle
   1997   Model-based testing. *Proceedings of the 10th International Software Quality Week '97*. San Francisco, CA: Software Research Institute.

Archer, M.
   1999   *Tools for simplifying proofs of properties of timed automata: The TAME template, theories, and strategies*. Technical report, NRL/MR/5540-99-8359. Washington, DC: Naval Research Laboratory.

Avritzer, A., and E.J. Weyuker
   1995   The automatic generation of load test suites and the assessment of the resulting software. *IEEE Transactions on Software Engineering* 21(9):705-716.

Baik, J., B. Boehm, and B.M. Steece
   2002   Disaggregating and calibrating the case tool variable in COCOMO II. *IEEE Transactions on Software Engineering* 28(11):1009-1022.

Bartoussek, W., and D.L. Parnas
   1978   Using assertions about traces to write abstract specifications for software modules. In *Proceedings of the Second Conference on European Cooperation in Informatics and Information Systems Methodology*, pp. 211-236. Venice, Italy.

Boehm, B., C. Abts, A.W. Brown, S. Chulani, B. Clark, E. Horowitz, R. Madachy, D. Reifer, and B. Steece
   2000   *Software Cost Estimation with COCOMO II*. Englewood Cliffs, NJ: Prentice Hall.

Chulani, S., B. Boehm, C. Abts, J. Baik, A. Windsor Brown, B. Clark, E. Horowitz, R. Madachy, D. Reifer, and B. Steece
   2000   Future trends, implications in cost estimation models. *Crosstalk, The Journal of Defense Engineering* 13(4):4-8.

Cohen, D.M., S.R. Dalal, A. Kajla, and G.C. Patton
 1994 The automatic efficient test generator (AETG) system. *Proceedings of the IEEE International Symposium on Software Reliability Engineering*, pp. 303-309.

Cohen, D.M., S.R. Dalal, J. Parelius, and G.C. Patton
 1996 The combinatorial design approach to automatic test generation. *IEEE Software* 13(5):83-88.

Dalal, S.R., and C.L. Mallows
 1988 When should one stop testing software? *Journal of the American Statistical Association* 83(403):872-879.

Dalal, S.R., and A.A. McIntosh
 1994 When to stop testing for large software systems with changing code. *IEEE Transactions on Software Engineering* 20(4):318-323.

Dalal, S.R., A. Jain, N. Karunithi, J.M. Leaton, and C.M. Lott
 1998 Model-based testing of a highly programmable system. *Proceedings of the ISSRE*, pp. 174-178.

Dalal, S., M. Hamada, P. Matthews, and P. Gardner
 1999 Using defect patterns to uncover opportunities for improvement. *Proceedings of the International Conference on Applications of Software Measurement (ASM)*, San Jose, CA.

Dalal, S.R., Y. Ho, A. Jain, and A. McIntosh
 2002 Application Performance Assurance via Post-Production Monitoring. Paper presented at the International Performance and Dependability Symposium, Washington, DC.

Faulk, S.R.
 1995 *Software Requirements: A Tutorial.* NRL Report 7775, Washington, DC: Naval Research Laboratory.

Ferguson, J.
 2001 Crouching dragon, hidden software: Software in DoD weapon systems. *IEEE Software* 18(4):105-107.

Garg, S., A. van Moorsel, K. Vaidyanathan, and K.S. Trivedi
 1998 A methodology for detection and estimation of software aging. *Proceedings of the Ninth International Symposium on Software Reliability Engineering*, Paderborn, Germany, pp. 282-292.

Gross, J., and J. Yellen
 1998 *Graph Theory and Its Applications.* Boca Raton, FL: CRC Press.

Harel, D.
 2001 From play in scenarios to code: An achievable dream. *Computer* 34(1):53-60.

Harel, D., and O. Kupferman
 2002 On object systems and behavioral inheritance. *IEEE Transactions on Software Engineering* 28(9):889-903.

Heitmeyer, C.L., R.D. Jeffords, and B. Labaw
 1996 Automated consistency checking of requirements specifications. *ACM Transactions on Software Engineering and Methodology* 5(3):231-261.

Heitmeyer, C., J. Kirby, B. Labaw, and R. Bharadwaj
    1998   SCR*: A toolset for specifying and analyzing software requirements. *Proceedings, Computer-Aided Verification, 10th Annual Conference* (CAV'98), Vancouver, Canada.

Heninger, K.
    1980   Specifying software requirements for complex systems: New techniques and their application. *IEEE Transactions on Software Engineering* SE-6(1):2-13.

Hester, S.D., D.L. Parnas, and D.F. Utter
    1981   Using documentation as a software design medium. *The Bell System Technical Journal* 60(8):1941-1977.

Holtzman, G.J.
    1997   The model checker SPIN. *IEEE Transactions on Software Engineering* 23(5):279-295.

Humphrey, W.S.
    1989   *Managing the Software Process*. Reading, MA: Addison-Wesley.

Jeffords, R., and C. Heitmeyer
    1998   Automatic generation of state invariants from requirements specifications. *Proceedings, 6th International Symposium on the Foundations of Software Engineering* (FSE-6), Orlando, FL.

Lee, D., and M. Yanakakis
    1992   On-line minimization of transition systems. In *Proceedings of the 24th Annual ACM Symposium on Theory of Computing,* pp. 264-274. New York: Association for Computing Machinery.

Leonard, E.I., and C.L. Heitmeyer
    2003   Program synthesis from formal requirements specifications using APTS, *Higher-Order and Symbolic Computation* 16(1-2):62-97.

Lithicum, D.
    2001   *B2B Application Integration: e-Business Enables Your Enterprise*. Reading, MA: Addison-Wesley.

Luckham, D.C., J.J. Kenney, L.M. Augustin, J. Vera, D. Bryan, and W. Mann
    1995   Specification and analysis of system architecture using rapide. *IEEE Transactions on Software Engineering* 4:336-355.

Lutz, R.R.
    1993   Analyzing software requirement errors in safety-critical, embedded systems. *Proceedings of the IEEE International Symposium on Requirements Engineering*, San Diego, CA.

Lyu, M.
    1996   *Handbook of Software Reliability Engineering*, Michael Lyu (ed.). New York: McGraw-Hill.

Meyers, S., and S. White
    1983   *Software Requirements Methodology and Tool Study for A6-E Technology Transfer*. Technical report. Bethpage, NY: Grumman Aerospace Corp.

REFERENCES

National Research Council
- 1996 *Statistical Software Engineering.* Panel on Statistical Methods in Software Engineering, Committee on Applied and Theoretical Statistics. Washington, DC: National Academy Press.
- 1998 *Statistics, Testing, and Defense Acquisition: New Approaches and Methodological Improvements.* Panel on Statistical Methods for Testing and Evaluating Defense Systems, Committee on National Statistics, Michael L. Cohen, John E. Rolph, and Duane L. Steffey, eds. Washington, DC: National Academy Press.
- 2002 *Reliability Issues for DoD Systems: Report of a Workshop.* Committee on National Statistics. Francisco Samaniego and Michael Cohen, eds. Washington, DC: The National Academies Press.

Oshana, R.S.
- 1999 An automated testing environment to support operational profiles of software intensive systems. *Proceedings of the International Software Quality Week '99.* San Francisco, CA: Software Research Institute.

Parnas, D.L., and Y. Wang
- 1989 *The Trace Assertion Method of Module Interface Specification.* Technical Report 89-261, Department of Computing and Information Science, Queen's University at Kingston, Kingston, Ontario, Canada.

Poore, J.H., and C.J. Trammell
- 1999 Application of statistical science to testing and evaluating software intensive systems, in *Statistics, Testing, and Defense Acquisition: Background Papers.* Panel on Statistical Methods for Testing and Evaluating Defense Systems, Committee on National Statistics, Michael L. Cohen, John E. Rolph, and Duane L. Steffey, eds. Washington, DC: National Academy Press.

Prowell, S.J., C.J. Trammell, R.C. Linger, and J.H. Poore
- 1999 *Cleanroom Software Engineering: Technology and Process.* Reading, MA: Addison Wesley.

Raffo, D., and M.I. Kellner
- 1999 Predicting the impact of potential process changes: A quantitative approach to process modeling. In *Elements of Software Process Assessment and Improvement*, Khaled El Emam and Nazim H. Madhavji, eds. New York: Wiley.

Rapps, S., and E. Weyuker
- 1985 Selecting software test data using data flow information. *IEEE Transactions on Software Engineering* SE-11(4):367-375.

Rosaria, S., and H. Robinson
- 2000 Applying models in your testing process. *Information and Software Technology* 42:815-824.

Umar, A.
- 2002 *E-business and Third Generation Distributed Systems.* St. Louis, MO: Academic Press.
- 2003 *Application Reengineering: Building Web-Based Applications and Dealing with Legacies.* Englewood Cliffs, NJ: Prentice Hall.

Vaidyanathan, K., and K.S. Trivedi
  1999 A measurement-based model for estimation of resource exhaustion in operational software systems. *Proceedings of the Tenth IEEE International Symposium on Software Reliability Engineering*, Boca Raton, FL, pp. 84-93.

Whittaker, J.A., and J.H. Poore
  1993 Markov analysis of software specifications. *ACM Transactions on Software Engineering and Methodology* 2(2):93-106.

Whittaker, J.A., and K. Agrawal
  1994 A case study in software reliability measurement. *Proceedings of the International Software Quality Week '94*, paper 2-A-2, San Francisco, CA.

Whittaker, J.A., and Michael G. Thomason
  1994 A Markov chain model for statistical software testing. *IEEE Transactions on Software Engineering* 20(10):812-824.

Woodcock, J., and J. Davies
  1996 *Using Z: Specification, Refinement, and Proof.* Englewood Cliffs, NJ: Prentice Hall.

APPENDIX A

Workshop Agenda and Speakers

AGENDA
Workshop on Statistical Methods
in Software Engineering for Defense Systems
July 19-20, 2001
National Academy of Sciences
Washington, DC

Thursday, July 19
Introduction/Overview
    Introductory remarks by Delores Etter, Office of the Undersecretary of Defense for Science and Technology
    Introductory remarks by Steven Whitehead, Operational Test and Evaluation Force
    Overview of workshop by Siddhartha Dalal, Telcordia Technologies; Jesse Poore, University of Tennessee

Model-Based Testing I
    Introductory remarks by Siddhartha Dalal, Telcordia Technologies
    Ashish Jain, Telcordia Technologies: "Functional Testing Using Combinatorial Designs"
    Manish Rathi, Telcordia Technologies: "AETG Web Service: A Smart Test Generation Service"

Jerry Huller, Raytheon: "Reducing Time to Market with Combinatorial Design Methods Testing"

Discussion: Linda Sheckler, Pennsylvania State University, Applied Research Laboratory
Discussion: David McClung, Applied Research Laboratories, University of Texas
Discussion: Luqi, Naval Postgraduate School

Model-Based Testing II
Introductory remarks by Jesse Poore, University of Tennessee
Stacy J. Prowell, University of Tennessee: "Testing Based on Markov Chain Usage Models"
Stacy J. Prowell, University of Tennessee: "Tool Support for Markov Chain Usage Model Testing"
Ron Manning, Raytheon: "Application"

Discussion: Mike Saboe, Tank Automotive and Armaments Command

Performance Analysis
Introductory remarks by Siddhartha Dalal, Telcordia Technologies
Elaine Weyuker, AT&T: "How to Use Testing to Measure Software Risk"
Kishor Trivedi, Duke University: "Modeling and Analysis of Software Aging and Rejuvenation"

Discussion: Brian Hicks, Air Force Materiel Command
Discussion: Lloyd Pickering, Army Evaluation Center

Data Analysis
Introductory remarks by Jesse Poore
Vic Basili, University of Maryland: "Using Defect Analysis to Identify Process Goals"
Barry Boehm, University of Southern California: "Bayesian Integration of Project Data and Expert Judgment in Parametric Software Estimation Models"
Ram Chillarege: "Orthogonal Defect Classification"

Discussion: Scott Lucero, Army SW Metrics, Army Evaluation Center
Discussion: William Farr, Naval Surface Warfare Center
Discussion: Patrick Carnes, Air Force Operational Test and Evaluation Center

Adjourn

Friday, July 20
Test Automation
Introductory remarks by Siddhartha Dalal, Telcordia Technologies
Harry Robinson, Microsoft: "Model-Based Testing in a Crunch-Time World"
Mike Houghtaling, IBM: "Aspects of Test Automation for Statistical Usage Certifications"

Discussion: David McClung, Applied Research Laboratories, University of Texas
Discussion: Ernie Seglie, Office of the Director of Operational Test and Evaluation

Design for Testability
Introductory remarks by Siddhartha Dalal, Telcordia Technologies
Jesse Poore, University of Tennessee: "Design for Testability"
Constance Heitmeyer, Naval Research Laboratory: "Methods That Improve the Requirements Specification"

Discussion: Tom Christian, Warner Robins Air Logistics Center
Discussion: Jack Ferguson, Office of the Undersecretary of Defense for Science and Technology

Integration Technology
Introductory remarks by Jesse Poore, University of Tennessee
Amjad Umar, Telcordia Technologies: "Integration Technologies: An Example-based Conceptual Overview"
Siddhartha Dalal, Telcordia Technologies: "Testing Integrated Applications and Services: A New Paradigm"

Discussion: Margaret Myers, Office of the Assistant Secretary of Defense for Command, Control, Communications and Intelligence

Discussion: Janet Gill, PAX River Naval Air Systems, Software Safety
Discussion: Stuart Knoll, Joint Staff, J-8

Closing Panel
Panel Members: Frank Apicella, Army Evaluation Center; Jim O'Bryon, Office of the Director of Operational Test and Evaluation; Steven Whitehead, Operational Test and Evaluation Force; Jack Ferguson, Office of the Undersecretary of Defense for Science and Technology; and Patrick Carnes, Air Force Operational Test and Evaluation Center

Closing Remarks:
Siddhartha Dalal, Telcordia Technologies; Jesse Poore, University of Tennessee

Tool Demos and Ad Hoc Discussion Groups

Adjourn

SPEAKERS WITH AFFILIATIONS
AS OF JULY 19-20, 2001

Frank Apicella, Technical Director, Army Evaluation Center
Vic Basili, Computer Science Department, University of Maryland
Barry Boehm, Center for Software Engineering, University of Southern California
Patrick Carnes, Air Force Operational Test and Evaluation Center, Test Support SW Division
Ram Chillarege, private consultant
Tom Christian, Warner Robins Air Logistics Center/Engineering
Siddhartha Dalal, Telcordia Technologies
Delores Etter, Office of the Undersecretary of Defense for Science and Technology
William Farr, Naval Surface Warfare Center/Dahlgren
Jack Ferguson, Office of the Undersecretary of Defense for Science and Technology, Director, Software Intensive Systems
Janet Gill, PAX River Naval Air Systems, Software Safety
Constance Heitmeyer, Head, Software Engineering Section, Naval Research Laboratory

Brian Hicks, Air Force Materiel Command
Mike Houghtaling, IBM Storage Systems Division
Jerry Huller, Raytheon
Ashish Jain, Telcordia Technologies
Stuart Knoll, Joint Staff, J-8
Scott Lucero, Army SW Metrics, AEC
Luqi, Naval Postgraduate School
Ronald Manning, Raytheon (TI Systems)
David McClung, Applied Research Laboratories, University of Texas
Margaret Myers, Office of the Assistant Secretary of Defense, Chief Information Officer, Command, Control, and Intelligence
James O'Bryon, Director of Operational Test and Evaluation
Lloyd Pickering, Army Evaluation Center
Jesse Poore, Department of Computer Science, University of Tennessee
Stacy J. Prowell, Department of Computer Science, University of Tennessee
Manish Rathi, Telcordia Technologies
Harry Robinson, Microsoft
Mike Saboe, U.S. Army Tank Automotive Research, Development
Ernest Seglie, Director of Operational Test and Evaluation
Linda Sheckler, Applied Research Laboratory, Penn State University
Kishor Trivedi, Department of Electrical Engineering, Duke University
Amjad Umar, Telcordia Technologies
Elaine Weyuker, AT&T
Steven Whitehead, Operational Test and Evaluation Force, Technical Advisor

APPENDIX
B

Glossary and Acronym List

INTEROPERABILITY TERMS

Application server: a collection of middleware services that have been packaged together for the development, management, and maintenance of end-user applications. Application servers can be for general purpose applications (e.g., IBM's WebSphere and Microsoft's .NET Framework) or for special purpose applications (e.g., Nokia's Application Server for Mobile Applications and Microsoft's E-Commerce Platform for e-commerce applications).

CORBA (common object request broker architecture): a middleware service that allows software application "objects" to interoperate over networks. For example a CORBA object representing inventory can interact with remotely located customer objects by using CORBA services.

Distributed database: a network of database servers that appear to users as a single system. Distributed databases address the problem of increasing demands for storage, sorting, and queuing as the quantity of information in a database becomes larger; for example a customer database can be distributed to the site where the customer resides in order to minimize network traffic.

DOM (document object module): a programming interface for XML/HTML documents that defines the structure of XML/HTML documents so that they can be accessed and manipulated by programs. For example, DOM allows a program (software agent) to read and understand the HTML/XML tags and act accordingly (e.g., search for customer names in an XML document containing customer names, addresses, etc.).

EAI (enterprise application integration platforms): platforms that permit existing applications to interact with each other in order to share business processes and data sources. Commercially available EAI platforms from suppliers (e.g., IBM, Tibco, Vitria) provide a "messaging bus" that can be used by diverse applications to communicate with each other.

EDI (electronic data interchange): an old (circa 1972) system for the exchange of information such as purchase orders, cataloguess, and invoices between organizations in a structured format. For many years, EDI has been used for business-to-business trade over privately owned corporate networks, but it is now competing against XML documents that are exchanged over the Internet and serve the same purpose.

EJB (Enterprise JavaBeans): This software, designed by Sun Microsystems, facilitates the development of middleware applications by providing support for services such as transactions, security, and database connectivity. EJBs, part of the J2EE specification (see below), use a "business component" approach where each application acts as a self-contained component that can be accessed and activated over the network.

JMS (JAVA message service): a flexible means for exchanging information between several clients and applications. It allows Java applications to exchange messages with each other asynchronously.

J2EE (JAVA 2 platform enterprise edition): a very comprehensive specification, defined by Sun for enterprise applications, for the infrastructure to support Web service and to enable development of secure and interoperable business applications. J2EE has several components, including JMS and EJBs.

Legacy wrapper: a software system that accepts requests from a "new" client (e.g., Web, CORBA) and adapts it for older (non-Web, non-CORBA) servers. These wrappers give legacy systems (e.g., an inventory system developed in the 1980s) newer "appearances" so that they can work with more recently developed applications. For this reason, these wrappers are also known as "face lifters."

Message broker: a software system that facilitates the integration of business applications, networks, and middleware processes through use of a hub and spoke–type architecture. Most EAI platforms defined above use message brokers at their core. Message brokers are similar to object request brokers (the core of CORBA) but are not restricted to exchanging object messages.

MOM (message-oriented middleware): a specific class of middleware that supports the exchange of general-purpose messages in a distributed application environment, i.e., it facilitates communication between distributed applications by using asynchronous processing. MOM is at the core of most message brokers.

ODBC/JDBC (open database connectivity/Java database connectivity): ODBC is an application programming interface for accessing relational databases, mainly used with C-based applications; JDBC is an application programming interface for accessing any tabular data source from Java.

PKI (public key infrastructure): the combination of software, encryption technologies, digital certificates, and other services that enables businesses to protect the security and integrity of their communications and transactions on the Internet.

SNA (systems network architecture): an old (circa 1972) proprietary network standard designed by IBM to handle communications and interactions between individual users on a networked system. Still in use in some industry segments, SNA is a proprietary network specification, as compared to the Internet, which is based on TCP/IP (an open network specification).

SOAP (simple object access protocol): a simple XML-based protocol designed to exchange information on the Web in order to provide Web services based on a shared and open Web infrastructure.

SSL (secure sockets layer): a method for protecting Web communications by providing data encryption, server authentication, and message integrity.

TP-Heavy and TP-Lite (transaction processing): two methods for database transactions: TP-Lite is limited to simple transactions with stored procedures, whereas TP-Heavy can monitor the execution of functions and permits interaction with other transactional units in a larger transaction.

WAP (wireless application protocol): an application environment and communication protocols for wireless devices to provide independent access to the Internet.

FAULT-TOLERANT SOFTWARE TERMS

Data diversity: a method for developing alternative software outputs from a single version of code through use of multiple, slightly perturbed versions of the input data; an example is n-copy programming.

Design diversity (parallel coding): for checking the accuracy of code, the use of sequentially or simultaneously available alternative versions of software written to the same specifications; examples include n-version programming and recovery block.

Environmental diversity (proactive fault management): a generalization of restarting a software system, and using instead a new or modified operating environment to enable the running software to avoid failure.

APPENDIX C

Biographical Sketches

DARYL PREGIBON *(Chair)* is division manager of the Statistics Research Department at AT&T Labs. His department is responsible for developing a theoretical and computational foundation for statistical analysis of very large data sets. He has nurtured successful interactions throughout AT&T, in fiber and microelectronics manufacturing, network reliability, customer satisfaction, fraud detection, targeted marketing, and regulatory statistics. His research contributions have changed from mathematical statistics to computational statistics, and included such topics as expert systems for data analysis, data visualization, application-specific data structures for statistics, and large-scale data analysis. He is a world leader in data mining, which he defines as an interdisciplinary field combining statistics, artificial intelligence, and database research. He received his Ph.D. in statistics at the University of Toronto (1979). He is a fellow of the American Statistical Association, former head of the Committee on Applied and Theoretical Statistics (1996-1998), and a member of the Committee on National Statistics (2000-present).

BARRY BOEHM is TRW professor of software engineering, in the Computer Science Department of the University of Southern California (USC), and director, USC Center for Software Engineering. He received his B.A. degree from Harvard in 1957, and his M.S. and Ph.D. degrees from UCLA in 1961 and 1964, all in mathematics. Between 1989 and 1992, he served within the U.S. Department of Defense as Director of the

Defense Advanced Research Projects Agency Information Science and Technology Office, and as Director of the DDR&E Software and Computing Technology Office. He worked at TRW from 1973 to 1989, culminating as chief scientist of the Defense Systems Group, and at the RAND Corporation from 1959 to 1973, culminating as head of the Information Sciences Department. He was a programmer-analyst at General Dynamics between 1955 and 1959. His current research interests include software process modeling, software requirements engineering, software architectures, software metrics and cost models, software engineering environments, and knowledge-based software engineering. His contributions to the field include the Constructive Cost Model (COCOMO), the Spiral Model of the software process, the Theory W (win-win) approach to software management and requirements determination, and two advanced software engineering environments: the TRW Software Productivity System and Quantum Leap Environment. He has served on the board of several scientific journals, including the *IEEE Transactions on Software Engineering, IEEE Computer, IEEE Software, ACM Computing Reviews, Automated Software Engineering, Software Process,* and *Information and Software Technology.* He has served as chair of the AIAA Technical Committee on Computer Systems, chair of the IEEE Technical Committee on Software Engineering, and as a member of the Governing Board of the IEEE Computer Society. He currently serves as chair of the Air Force Scientific Advisory Board's Information Technology Panel, and chair of the Board of Visitors for the CMU Software Engineering Institute. He is an AIAA Fellow, an ACM Fellow, an IEEE Fellow, and a member of the National Academy of Engineering.

MICHAEL L. COHEN is a senior program officer for the Committee on National Statistics. Previously, he was a mathematical statistician at the Energy Information Administration, an assistant professor in the School of Public Affairs at the University of Maryland, a research associate at the Committee on National Statistics, and a visiting lecturer at the Department of Statistics, Princeton University. His general area of research is the use of statistics in public policy, with particular interest in census undercount, model validation, and defense testing. He is also interested in robust estimation. He has a B.S. degree in mathematics from the University of Michigan and M.S. and Ph.D. degrees in statistics from Stanford University.

SIDDHARTHA R. DALAL is vice president in the Xerox Innovation Group at the Xerox Corporation in Webster, NY. He previously created and managed a multidisciplinary research division concentrating on amalgamation of Information Services (e-commerce, multimedia, mobility, telephony, software engineering, etc.) and Information Analysis (data mining/statistics, knowledge management) at Telcordia Technologies. He initiated and led research that resulted in commercial products for Telcordia's Software Systems, consulting ware for Professional Services, and initiated new research areas in software engineering, data mining, and marketing areas. He started and promoted research programs that reduced costs and created new lines of business. Examples include Year 2000 Testing, Software Process, first ASP service at Telcordia (AETGWeb), Software Testing (When to Stop Testing, Test Case Generation), Risk Analysis, and Data Mining. Recent projects include research on e-commerce in the telecom marketplace, testing (test automation, GUI, and Web testing), rule-based systems and agents, assessment methodologies for software processes (like CMM and SPICE), defect analysis, and a prediction tool for market penetration of new products. He received the Telcordia CEO award (1999), he is an elected member of the International Statistical Institute, a Fellow of the American Statistical Association, a member of the Committee on Applied and Theoretical Statistics (CATS). He received his MBA and his Ph.D. in statistics from the University of Rochester.

WILLIAM F. EDDY is a professor in the Department of Statistics, Carnegie Mellon University. His research interests include data mining, dynamic data visualization, and generally computational and graphical statistics. He is a fellow of the American Association for the Advancement of Science, fellow of the American Statistical Association, founding editor of the *Journal of Computational and Graphical Statistics*, fellow of the Institute of Mathematical Statistics, elected member of the International Statistical Institute, and fellow of the Royal Statistical Society. He is founding co-editor of *Chance*. He has served on the following panels and committees of the National Research Council: Commission on Physical Sciences, Mathematics, and Applications (CPSMA) Working Group on Infrastructure Issues for Computational Science (1992-1993), CPSMA Panel on Statistics and Oceanography (co-chair, 1992-1993), CPSMA Board on Mathematical Sciences (1990-1993), Committee on Applied and Theoretical Statistics (CATS) Panel on Guidelines for Statistical Software (chair, 1990-1991), CATS (1987-1993; chair 1990-

1993), CNSTAT Panel on Statistics on Natural Gas (1983-1985), and CNSTAT (1994-2000).

JESSE H. POORE holds the Ericsson/Harlan D. Mills Chair in Software Engineering in the Department of Computer Science at the University of Tennessee, and is Director of the University of Tennessee-Oak Ridge National Labs Science Alliance, which is a program to promote and stimulate joint research between the University of Tennessee and Oak Ridge National Labs, and to manage joint programs and encourage interdisciplinary collaborations. He conducts research in cleanroom software engineering and teaches software engineering courses. He has held academic appointments at Florida State University and Georgia Tech; he has served as a National Science Foundation rotator, worked in the Executive Office of the President, and was executive director of the Commitee on Science and Technology in the U.S. House of Representatives. He is a member of ACM, IEEE, and a fellow of the American Association for the Advancement of Science. He holds a Ph.D. in information and computer science from Georgia Tech. He also served on the Committee on National Statistics' Panel on Statistical Methods for Testing and Evaluating Defense Systems (1994-1998).

JOHN E. ROLPH is professor of statistics and former chair of the Department of Information and Operations Management in the University of Southern California's Marshall School of Business. He previously was on the research staff of the RAND Corporation and has held faculty positions at University College London, Columbia University, the RAND Graduate School for Policy Studies, and the Health Policy Center of RAND/University of California, Los Angeles. His research interests include empirical Bayes methods and the application of statistics to the law and to public policy. He has served as editor of the American Statistical Association magazine *Chance,* and he currently is chair of the National Research Council's Committee on National Statistics (CNSTAT). He also served as chair on CNSTAT's Panel on Statitistical Methods for Testing and Evaluating Defense Systems. He is a fellow of the American Statistical Association, the Institute of Mathematical Statistics, the American Association for the Advancement of Science, and is an elected member of the International Statistical Institute. He received A.B. and Ph.D. degrees in statistics from the University of California at Berkeley.

FRANCISCO J. SAMANIEGO is professor of the Intercollege Division of Statistics and Director of the Teaching and Resources Center at the University of California at Davis. He has held visiting appointments in the Department of Statistics at Florida State University and in the Department of Biostatistics at the University of Washington. His research interests include mathematical statistics, decision theory, reliability theory and survival analysis, and statistical applications, primarily in the fields of education, engineering, and public health. He is a fellow of the American Statistical Association, the Institute of Mathematical Statistics, and the Royal Statistical Society. He is an elected member of the International Statistical Institute. He received a B.S. degree from Loyola University of Los Angeles, an M.S. degree from Ohio State University, and a Ph.D. from the University of California, Los Angeles, all in mathematics. He served as a member of the CNSTAT Panel on Statistical Methods for Testing and Evaluating Defense Systems (1994-1998), and as a member of CNSTAT (1995-2001). He also chaired the first workshop in a series of workshops on statistical methods relevant to the development of defense systems, of which this workshop is the second.

ELAINE WEYUKER is a technology leader at AT&T Laboratories in Florham Park, NJ, and was recently selected to be an AT&T Fellow. Before moving to AT&T Labs in 1993, she was a professor of computer science at the Courant Institute of Mathematical Sciences of New York University where she had been on the faculty since 1977. Prior to that she was a faculty member at the City University of New York, and was a systems engineer at IBM and a programmer at Texaco, Inc. Her research interests are in software engineering, particularly software testing and reliability, and software metrics. She has published more than 100 refereed papers in journals and conference proceedings in those areas, and has been a frequent keynote speaker at software engineering conferences. Many of her recent publications involve large case studies done with industrial projects assessing the usefulness of proposed new testing strategies. She is also interested in the theory of computation, and is the author of a book (with Martin Davis and Ron Sigal), *Computability, Complexity, and Languages, 2nd Ed.* In addition to being an AT&T Fellow, she is also a fellow of the ACM and a senior member of the IEEE. She was recently elected to the Board of Directors of the Computing Research Association, and is on the Technical Advisory Board of Cigital Corporation. She is a member of the editorial boards of *ACM Transactions on Software Engineering and Methodology*, the

Empirical Software Engineering Journal, and is an advisory editor of the *Journal of Systems and Software.* She has been the Secretary/Treasurer of ACM SIGSOFT and was an ACM National Lecturer. She received a Ph.D. in computer science from Rutgers University, and an M.S.E. from the Moore School of Electrical Engineering, University of Pennsylvania.